For You

Andreas Seidl

Handover of Power

European Version

Volume 13: Innovation

Imprint

Bibliographic information of the German National Library:
The German National Library lists this publication in the
German National Bibliography; detailed bibliographic data
are available on the Internet at http://dnb.dnb.de.

© 2022 Dipl. Pol. Theodor Andreas Seidl

Cover: Christiane Ebrecht
Translation: DeepL, Cologne
Production and publishing: BoD – Books on Demand,
Norderstedt

ISBN: 978-3-7568-0262-3

Acknowledgements

My thanks go to my family and friends who have made me who I am today. Special thanks to all those who supported me in writing this book. I would like to thank all my classmates, teachers, fellow students, lecturers, demonstrators, activists, colleagues, companies and countries with whom I have had the privilege of sharing the experiences from which all the ideas in this book have emerged. I would like to thank the staff of Books on Demand for their kind helpfulness. I thank the citizens of Seligenstadt for the harmony and solidarity in which I was able to write.

Foreword

This policy concept contains a variety of proposals for possible political reforms. It can be peacefully and democratically adapted to any current political system of any state in the world, but also to political systems in families, clubs, associations or companies. Wherever humans make or submit to rules that manage living together, the following proposals can be helpful. Readers who find the proposals so helpful that they would like to implement them together with like-minded people can contact the author. The contact form on the last page can be used for this purpose.

Faults and defects

I ask for your understanding that this volume was not professionally proofread. I could only afford professional proofreading for the summary. Spelling errors and unfortunate phrasing may therefore occur. As soon as this volume has sold enough to pay for a professional proofreading, it will be done. After that, a new edition will be published.

English version

Please understand that this volume has been translated automatically. I could only afford a professional translation for the summary. Poor wording and spelling errors may therefore occur. In case of doubt, the German version shall prevail. As soon as this volume has sold enough to pay for a professional translation, it will be done. After that, a new edition will be published. It was more important to me that no one in the world should have an information advantage than individual translation errors in the complete work.

References
If something has been quoted directly, it is set in italics. If the headings contain footnotes, the sources for direct and indirect quotations apply in the chapter for which the heading stands. Otherwise, quotations or source references are directly at the word or at the end of the sentence or paragraph. This book contains parts of text based on the Federal Constitution of the Swiss Confederation of 18 April 1999 (as of 12 February 2017), abbreviated to BV[1] and the Constitution of the Canton of Bern of 6 June 1993 (as of 11 March 2015), abbreviated to KV[2] .

If the constitutional paragraph, or individual paragraphs thereof, are based in whole or in part on extracts from the BV or KV, this is indicated in a footnote. The references to the corresponding footnotes for constitutional paragraphs are usually found after the heading of the affected chapter and sometimes in the body of the text. Articles used in the Swiss constitutions are listed in the footnote with a number after the title of the constitutional paragraph. Example: §123 Sample title: BV Art.123, KV Art.123.

All internet sources are fully cited in the footnotes. They were last accessed on 30.09.2021. All literature sources are also listed in full in the footnotes.

All references to tasks undertaken by other ministries and described in more detail there are given in footnotes. Example: Model Ministry - 1.2.3 Model Chapter.

All footnotes are to be viewed in comparison to the respective source, so-called indirect quotations. Direct quotations are set in italics, but hardly ever occur. The source reference is intended to enable further investigation and to take copyright

---

1 This is not an official publication. Only the publication by the Swiss Federal Chancellery is authoritative. https://www.fedlex.admin.ch/eli/cc/1999/404/de On 14.12.2021

2 This is not an official publication. The Bernese Official Collection of Laws is authoritative. https://www.belex.sites.be.ch/frontend/versions/2420?locale=de#ART71 On 16.12.2021

into account.

All keywords used, based on the names of the responsible units, departments and ministries of Germany, are listed at the end of this volume in the chapter on the conversion of ministries.

Table of contents

1 Goals of the Ministry of Innovation

The Ministry of Innovation aims to produce and market as many new inventions as quickly as possible. In this way, it contributes to raising the standard of living of humankind and to the accumulation of wealth for inventors and savers, as well as to the reduction of taxes through patented inventions marketed by the state.

Ideas that increase technical and social progress are the source of all growth in living standards. The growth rates of this innovative economic activity must be used in the sense of economic entrepreneurship and maintained in the long term. In the information age, knowledge is experiencing a doubling time that is getting shorter and shorter. Development must keep up with this rate of research and is therefore promoted by this ministry. There are no limits to growth through intelligence, unlike the maximum number of population or consumers. Intelligence is a renewable resource that can be coupled using digital technology and thus accumulated. Knowledge is the basis for making the raw material of intelligence usable. Fortunately, knowledge can be shared as often as desired and only has high fixed costs when it is created, but the transmission of this knowledge has significantly lower variable costs when it is shared. In the process, shared knowledge creates new value when it is shared, in that humans often get ideas about how to deal with something when they learn it as playfully as possible.

Knowledge sharing uses digital information and communication technology. Computing power is increasing and the price of it is falling all the time. Through the digital technologies that are in computers, human brains can more easily network to share knowledge. Through this technology, a people can make a problem public and work together to find a solution. Companies can offer market-ready solutions and sell them on the market. The state can offer majority-ready solutions and ask voters to cast their ballots. The Ministry of Innovation offers this digital networking with the challenge search engine and the directories for research and ideas.

Because today's population is intelligent and educated, scientific and technical challenges can be addressed publicly to arouse eagerness to master them. Whoever is the first has an advantage on the market through industrial property rights, and this is how further development is created. Once a human has mastered the challenge, the remaining humans can concentrate on other challenges. Once the period for an industrial property right has expired, these inventions then become accessible to the whole world free of charge.

In an increasingly automated world of work, it is possible for humans to take on work that computers cannot, namely creative innovative thinking and mutual inspiration. In order to derive new tasks for machines from this and to profit from their value creation, the Ministry of Innovation offers various employment opportunities for inventors. This is becoming increasingly important as the Unconditional Basic Income[1] rises. Inventions are a possible source of additional income in these times.

2 Departments

The departments are divided into sub-departments and enumerations are usually considered as their individual units. Many tasks of some departments are completely taken over by other ministries as a service.

2.1 Central Department

Part of the Central Department is the Reception Office with the Courier and Mail Room, which directs all concerns, broadcasts and visitors to the appropriate place in the ministry.

2.1.1 Staff

The Human Resources Department is responsible for staff development and planning. For this purpose, it takes care of the recruitment of junior staff, intern and trainee programmes as well as the selection procedures for employees and special selection procedures for applicants with disabilities. For politicians and employees, the department prepares a job plan.

---

1 Ministry of Finance - 6 Unconditional Basic Income

In all its tasks, it works in voting with the personnel board.[2] All other personnel matters are transferred to the relevant ministries. The Ministry of Education is responsible for the training and further education of employees for the state service.[3] The Ministry of Labour takes over the service law.[4] This includes the labour and collective bargaining law for employees in the state service, remuneration, personnel administration of all careers and employees, flexitime, holiday and sickness records, working time with or without flexitime in part-time or full-time at the place of work or in home work. The Ministry of Infrastructure provides housing assistance for all state employees.[5] The Ministry of Finance's Pay Office takes care of employees' salary, expenses, travel and relocation costs.[6]

The Ministry of Education provides childcare for all employees in the state service.[7]

The Ministry of Health is responsible for the occupational health service.[8] It ensures occupational health management, deals with the treatment, education and prevention of occupational accidents, controls and provides occupational health and safety through the health auditors[9] of the Company Auditing Agency[10] .

2.1.2 Organisation

The ministries of media, security, justice, finance, labour, state organisation provide audit services for quality management in the ministry, evaluation of work performance, revenues and expenditures, as well as corruption prevention, sabotage

---

2 Ministry of State Organisation - 2.1.1.1 Personnel board
3 Ministry of Education - 2.1.1.1 Education and training for the state service
4 Ministry of Labour - 4 State enterprises, 13 Labour Directory
5 Ministry of Infrastructure - 2.1.1.1 Housing assistance for state service employees
6 Ministry of Finance - 2.1.1.1 Staff remuneration
7 Ministry of Education - 2.1.1.2 Childcare for state service employees
8 Ministry of Health - 2.1.1.1 Occupational Health Service
9 Ministry of Labour - 20.7.2 Health auditor
10 Ministry of Labor - 20 Company Auditing Agency

protection and, if necessary, disciplinary matters.[11]

The language service for translating talks or texts is provided by the Ministry of Education.[12] The Ministry of Finance organises the annual budget vote and ensures proper accounting in each ministry.[13] It regulates budget procedures, budget law, staff budgets, departmental budgets, costs and cash management, and assists ministries in budget planning for the budget vote. The Ministry of Labour regulates procurement law and ensures corruption-free state orders and procurement.[14]

The Ministry of Digital Affairs supports the supply of Information Technology.[15] In voting with the Procurement Office of the Ministry of Labour, it takes care of the procurement, provision, maintenance and service of technical devices and software. Much of this is produced in-house to ensure data protection in information and communication technology. Information technology and digitalisation officers audit and advise the ministries. Digital appointment calendar and documentation services are provided as well as a digital policy archive including a library.

2.2 Management Department

The Management Department is the minister's department. With his office team, he provides policy planning and analysis for his ministry and coordinates the relationship between the nation and the municipality through exchanges with his deputies in the municipalities. He initiates cooperation with other ministries or citizens in committees and is supported by the Ministry of State Organisation.

The Ministry of Media Affairs, through its media service, provides press and public relations for the ministry, moderates civil dialogue, trains or provides a spokesperson for the minister, writes speeches and texts on request, and ensures the

---

11 Ministries of Media, Security, Justice, Finance, State Organisation - 2.1.2.1 Audit services
12 Ministry of Education - 2.1.3 Language Service
13 Ministry of Finance - 8 state revenues, 9 state expediture
14 Ministry of Labour - 6 Procurement Office
15 Ministry of Digital Affairs - 2.1.2.1.1 Supply of Information Technology

implementation of conferences and events.[16]

The Ministry of Digital Affairs is responsible for digital management and thus provides departmental management. It automatically produces business statistics, staff surveys and the current state of research through statistics. It automatically forwards proposals to the affected or empowered state employees. In document management, it ensures digitalisation and that ministries share forms with each other.[17]

2.3 European Department

The Ministry of Foreign Affairs ensures the constant transmission of the latest information on current European policy affecting the ministry concerned, applicable European Union law and all European Union funding programmes starting or in progress.[18]

This department is responsible for the European funding programmes for research and development[19] . It prepares the programmes in voting with the Minister for Innovation in order to negotiate them with all ministers for innovation in the European Union. If the funding programmes are effective, this department works with the indoor service of the Innovation Agency to ensure that as many innovative companies and inventors as possible can benefit from the funding. The innovation auditors and the field service contact suitable companies and inventors.

The European Department decides for the areas of research and innovation[20] whether current European Union law is adopted, adapted or rejected.[21]

---

16 Ministry of Media Affairs - 2.2.1.1 Media Service
17 Ministry of Digital Affairs - 2.1.2.1 Digital Service
18 Ministry of Foreign Affairs - 2.4 European Department
19 https://www.bmwi.de/Redaktion/DE/Artikel/Mittelstand/erp-sondervermoegen.html https://www.bmbf.de/de/horizont-2020-das-europaeische-forschungsrahmenprogramm-281.html https://ec.europa.eu/esf/home.jsp?langId=de
20 https://eur-lex.europa.eu/summary/chapter/research_innovation.html?root_default=SUM_1_CODED=27
21 Ministry of Foreign Affairs - 6.4 Conversion of political contents to the policy of dynamic media democracy

2.4 Technology Policy Department

The Technology Policy Department provides oversight for the Patent Office, the Innovation Agency and the existing People's Innovation Companies. This department primarily supervises other states, ministries, own and foreign authorities. It organises cooperation between municipalities so that similarities in different locations can be identified and contact can be established, and the division of labour and tasks can be coordinated across the country. It ensures a regionally equivalent offer of innovation promotion and the same standards. As soon as other ministries become involved, this department provides mediation wherever the responsibility of the Ministry of State Organisation ends, namely after the legislation has been passed, when it comes to the execution of the laws. Whenever the Ministry of Foreign Affairs is involved, this department takes over contact with innovation ministries of other states.

2.5 Innovation Policy Department

The Innovation Policy Department ensures the promotion of research, development, invention, financing and commercialisation of innovations. This department primarily serves humans who want to drive innovation in a research, inventive or entrepreneurial way. To this end, it ensures the operation of the directories for research and ideas in voting with the Ministry of Digital Affairs, as well as the operation of the Ideas Stock Exchange in voting with the Ministry of Finance. It ensures that all measures to promote innovation are utilised as comprehensively as possible and can draw up further measures in voting with affected parties and propose them to the Minister of Innovation or terminate unnecessary measures. It supervises state research projects and ensures sufficient freedom and fairness in inventive activities in state institutions. It oversees the measures on symbol policy and manages related events and selection procedures.

3 Tasks of the Ministry of Innovation

The ministry's task is to ensure that there are more innovations in the country and that they are marketed faster and more successfully. To ensure more innovation, the ministry promotes networking among research institutions, inventive activity and networking between companies and state research and educational institutions. It also awards state research projects in voting with the people. In order to be able to market innovations more quickly, the process from finding an idea to granting a suitable industrial property right to marketing is accelerated with promotional measures and a regulated schedule. In order to market innovations more successfully, the Ministry provides contact opportunities between researchers, inventors, entrepreneurs and investors via the Research Directory[22] , the Ideas Directory, the Ideas Stock Exchange and the Innovation Office. Highly promising innovations registered with the Patent Office or in the Ideas Directory can be connected by the Ministry of Innovation with an offer to operate a People's Innovation Company directly when the industrial property right is granted. People's Innovation Companies are the Ministry of Innovation's means of implementing urgently needed innovations and replacing taxpayers' money with profits.

The Innovation Agency's staff accompany and support researchers, inventors, entrepreneurs and investors in the introduction of innovations and difficulties in their implementation. They are supported by searchers and auditors from the Patent Office and by auditors and advisors from the Company Auditing Agency.

The Ministry of Innovation ensures networking, cooperation and voting with the ministries of labour, economy, education, health, infrastructure and foreign affairs in advancing innovations and research projects. When setting priorities, it pays attention to sustainability for the environment and future generations. Because innovation is cyclical, the Ministry of Innovation has the task of effectively accompanying the current cycle and preparing the entrance into the next cycle and coordinating it with other countries around the world.

22Ministry of Digital - 12 Directories

3.1 Innovation cycles

According to the So-called Kondsratyev cycle[23] , humanity is currently in the Information Technology cycle. Upcoming cycles are always associated with a certain probability and are actually only confirmed in review when society is already in the middle of the cycle. Upcoming cycles could be biotechnology, nanotechnology, robotics, energy from renewable sources or nuclear fusion, psychosocial health or resource efficiency.

In the short term, humanity is in the fifth innovation cycle of Information Technology, culminating in the introduction of the intranet and the digital direct-democratic management of states by their citizens. In the medium term, biotechnology will determine the sixth cycle and lead to a globally self-sufficient supply. Natural ecosystems will then be replicated by humans with animal and plant species useful to them, which will provide sufficient food and building materials. Robotics will be expanded to include bionics and parts will grow as desired in a nutrient solution. Biotechnology makes it possible to transform machines into living beings made of renewable and biodegradable biomass instead of rare precious metals and toxins. The economic cycles are then completely closed, every waste is recycled, as in natural ecosystems. In the long term, i.e. in the seventh innovation cycle, space technology is in vogue because a second Earth has been found that needs to be colonised.

3.2 International innovation policy

National and international innovation and technology policy is coordinated by the indoor service of the Innovation Agency. The goal is to create an international division of labour in research areas and to agree on a current and upcoming innovation cycle that will be jointly advanced. The Ministry of Innovation has the task of agreeing on a direction in voting with European Union member states and other countries.

In cooperation with the Ministry of Foreign Affairs, biotechnology is to be promoted worldwide if possible. It draws

---

23https://de.wikipedia.org/wiki/Kondsratjew-Zyklus

its resources from renewable sources, produces biodegradable waste and can solve tasks in robotics, nanotechnology and energy production through genetic manipulation. The peoples of as many countries as possible determine in an ethics committee[24] how far genetic engineering may go in order to create intelligent life. Each participating country can choose several animal and plant species so that not many study the same species, although there are millions of useful species. For example, Spain and Italy could specialise in insects. Spain studies the muscles of flying insects and Italy studies the exoskeleton made of chitin of swimming and diving insects. The two southern countries are taking on this task because insects in warmer regions can be better bred en masse to produce raw materials or materials. Each species or genus is to be researched as far as possible so that all processes can be understood and replicated by humans under laboratory conditions.

4 Innovation Agency

The Innovation Agency ensures that everything runs smoothly from the development of an idea to its marketing and the expiry of the industrial property right. It is divided into an indoor service and a field service.

4.1 Indoor service

The indoor service is located in the capital city of the Ministry of Innovation. It ensures cooperation between the institutions of the Ministry of Innovation and with institutions of other ministries. The institutions of the Ministry of Innovation are the Innovation Offices, the Patent Office and People's Innovation Company. The institutions of other ministries are educational institutions[25] , the Company Auditing Agency[26] ,

---

24 Ministry of State Organisation - 8.5.9 Ethics Commission
25 Ministry of Education - 4.10 Education through research
26 Ministry of Labour - 20.7 Company Auditing Agency departments

Party Television[27] , the Construction Team[28] and the Ideas Stock Exchange[29] . With the educational institutions, innovation is done through education. The innovation auditors of the Company Auditing Agency certify and market innovations, the technical auditors test new technologies and the business consultants support innovative business start-ups. The Party Television[30] produces promotional videos for innovations and marketing shows. The Construction Team builds factories for People's Innovation Company and the Ideas Stock Exchange links innovations with investors.

4.2 Field service

The field service maintains the innovation offices in the town halls. For inventors, this agency with its field service is the direct point of contact from the discovery of the idea to the end of the term of validity of the industrial property right. The field staff ensure that inventors are accompanied as early as the idea stage and offer consultation hours for inventors in the Innovation Office. Field staff are bound to professional secrecy. They advise inventors on which industrial property right is suitable for their idea and establish contact with companies or institutions of other ministries.

4.2.1 Innovation Office

In the Innovation Office, inventors, companies and investors can receive all the services that are also available through the Ideas Directory. This is intended to enable analogous humans to market their ideas as well. Users of the Ideas Directory can receive advice or support at the Innovation Office if they have questions or are uncertain about individual steps in the registration or marketing process. The innovation offices are located in every town hall.

---

27 Ministry of Media Affairs - 2.2.1.1 Media Service, 10.1.3.4 Crowdfunding
28 Ministry of Infrastructure - 5.8 Construction Team
29 Ministry of Finance - 11.9 Ideas Stock Exchange
30 Ministry of Media - 10 Party Television

4.2.1.1 Investigation

The Innovation Office can be asked for advice on which search command is needed to get an apt investigation into whether someone already had the idea. The hit list of this search query can be sent to the inventor so that he or she can search the hit list at home or in the intranet café. Ideas Directory investigations can be done on the inventor's own People's Computer or in the Intranet Café. International investigations must be carried out via the Internet. The inventor receives the search commands necessary for a suitable hit list in an email.

4.2.1.2 Analogue login

Inventors can check an idea for protectability in the Innovation Office together with the field staff member and register all possible industrial property rights. The field staff member creates a new invention profile in the name of the inventor in the Ideas Directory. For this purpose, the identity card is needed to link the new invention profile with the existing profile of the inventor(s) in the Persons Directory.
If there are several inventors, an inventor group must be formed and all participants must give their consent in person or digitally via the Ideas Directory. If inventors have already used the video log function, the video is saved in the invention profile. The video will be fast-forwarded to screen the involved inventors and check their identity against their identity cards. Inventors bring their invention to the Innovation Office and photos are taken or a file, text, drawing on paper or other storage medium is brought and viewed. To examine inventions that are not transportable, the field service may travel to the inventions. A visual inspection follows, but no substantive examination. All data is uploaded to the invention profile of the Ideas Directory.

4.2.1.3 Marketing

Inventors can have their industrial property right marketed through the Innovation Office or receive advice on the various marketing options offered by the Ministry of Innovation. The advice is free of charge; when marketing through the Innovation Office, the Ministry of Innovation receives a profit share of 20% before tax.

5 Research and development[31]

The Ministry of Innovation pursues a research policy that connects basic research, applied research and industrial research. Basic research mainly conducts research and hardly any development. While basic research exclusively brings to light new findings in all sciences because it wants to fully understand the environment, other subfields can put these findings to use in new technologies. Applied research carries out both basic research and development for practical use. Industrial research carries out little research and mainly development for commercial use.

All three sub-areas of research are to be researched by all research institutions. A research institution may specialise in one sub-area, but must make proposals for other sub-areas on how the research results can be used in the other sub-areas. State research institutions are obliged to conduct research in all three sub-areas. They also indicate which sub-area could make further use of their research results. In order to make further use possible in suitable research projects or suitable research institutions, an algorithm records all information and proposals in the Research Directory and finds suitable profiles and groups.

Nationals have the lifelong right to conduct research in the Innovation Labs and educational institutions. In doing so, they must partly adhere to the given requirements or adapt to conditions. Scientific freedom applies to all researchers. They are allowed to research what they want, but must not endanger humans or the environment in the process. Endangering

---

31 §17,2 Right to education and research, §18 Scientific freedom: BV Art.20, KV Art.21, §183,1 Research and innovation: BV Art. 64

animals and plants may be permissible for experimental purposes. The endangerment of humans may be permissible if the test subjects give their consent after being fully informed about the experiment. In case of doubt, the health auditors of the Company Auditing Agency decide.

5.1 European cooperation

In voting with the Ministry of Foreign Affairs, various fields of research are explored in the different member states of the European Union. All member states form a community of European research organisations to solve all urgent research tasks in a division of labour. For example, Germany could specialise in life sciences and France in economics, or Germany could research mobility and France energy sources.

5.2 Research institutions

Research institutions are mobile Innovation Labs, educational institutions, institutes and research departments of domestic citizens, which together form the Community of Domestic Research Institutions managed by the Innovation Agency. The indoor service promotes networking among the members of the community of domestic research institutions in the Research Directory and provides funding through the Research Cost Fund.

Research institutions are required to conduct basic research, to conduct only sustainable research, to share their premises or materials as long as they do not use them themselves and to network via the Research Directory.

Applied research and industrial research must always devote 10% of their annual research activities to basic research. It is possible to carry out a large-scale basic research project over a period of months using all available resources and then not have to do any basic research at all for many years.

Sustainable research means that no development may be carried out that leads to hazardous waste that cannot be disposed of or causes damage or late effects to humans and the environment. For example, the development of a nuclear power plant would be prohibited if radiated nuclear waste cannot be rendered harmless. Research, on the other hand, is permitted. For example, a nuclear power plant could be built for testing purposes to produce radioactive waste that can be used to develop procedures for disposal. Shifting the residue of research to subsequent generations through an economic use of dangerous research results is considered intentional bodily harm to subsequent generations.

All research institutions must enter all their research projects in the Research Directory, but can deny visibility to other users. Only the Innovation Agency has viewing rights, because it is subject to the duty of confidentiality and the innovation auditors use the algorithm to find similar or matching research projects.

5.2.1 Free researchers

The citizens who like to do research in their free time are called free researchers. In return, the independent researchers must work through a list of research tasks to approve the institution's professional researchers. Independent researchers should be given the opportunity to conduct basic research as well as to develop prototypes. Freelance researchers can also participate in a research community[32] and work through corresponding tasks and tests from a plan.

All research institutions report their available premises and equipment to the Innovation Agency. The field service ensures free admission to research opportunities for all citizens in voting with the research institutions. Research institutions open their premises to free research outside working hours. This is voluntary for private research institutions, mandatory for state research institutions. Ongoing research projects and secret research for which industrial property rights are to be registered or trade secrets are to be developed must not be hindered. The corresponding premises and objects are then not accessible.

32Ministry of Education - 11.7.1 Research community

5.3 Research Directory

In the Research Directory, all research projects are given a profile. The profiles are categorised according to basic research, applied research and industrial research and divided into subject areas and sectors. Companies that want to have something researched create a research assignment in an existing research project or a new research project for this purpose. Research communities in Innovation Labs, Think Tanks, colleges and institutes can take the research assignments and modify them in voting with the companies. Any research institution can post research projects as a profile and become the first member of the research group, which other research institutions can join. All researchers involved in a research project open a research group. Each individual research institution is given a sub-group. If tests and studies are needed for the research, the companies that would benefit from the research results can participate in the study. Planned Economy[33] and Social Market Economy[34] companies are required to participate in studies if they are selected by state colleges or institutes to do so.

Companies that have a need for research and development create a profile for a research project, describing only the outcome or goal, but not the research process. Research institutions can register with the appropriate group in the Research Directory to complete research orders for a research project. Via the profile of a research order, an application can be made directly to the Research Cost Fund for the assumption or subsidy of the costs.

5.3.1 Research Cost Fund[35]

The Research Cost Fund is a funding mechanism for each profile in the Research Directory. Each profile receives an account into which money can be deposited, a So-called pot. Each user can deposit money if they want to support the research project. Companies can invest part of their research expenditure in pots of matching research projects to drive a particular innovation that could give them more efficient production or more future-proof products or services. The pots can be open or closed. Companies can profile self-interested research projects to create a closed pot or select existing closed or open pots to participate in the deposit.

5.3.1.1 Closed pots

Closed pots can be established by domestic entrepreneurs when they create a new profile in the Research Directory. This allows them to share their research costs with other domestic companies to jointly fund research projects that benefit all participants. Since research costs for companies are often high risk and difficult to predict returns, they can pool their funds by investing in matching research projects. For example, a battery manufacturer can invest in the electricity pot.

Entrepreneurs can invite researchers, developers and inventors via their profile in the Labour Directory to use funds from the corresponding pots, from the results of which they expect further developments. The innovation auditors of the Company Auditing Agency[36] identify synergy effects, find similar or identical research projects and development challenges when auditing the companies. The companies receive a proposal in the audit results to pool their funds for research and development with other companies facing the same challenges. Closed pots can encourage graduates at educational institutions to write theses that address the desired research topic. The graduates receive the corresponding payments from the Research Cost Fund. The companies administer their closed pots independently and check to whom the money should be paid.

36Ministry of Labour - 20.7.5 Innovation auditor

5.3.1.2 Open pots

Open pots can be founded by any inventor or researcher. All companies that could benefit from the proposed research or development can invest in Open Pots. Open pots are suitable for freelance researchers, inventors and developers to raise money for research and development. Marketing is handled by the investing companies. Innovation auditors administer the open pots and decide on disbursements. As soon as a research project can demonstrate progress ready for series production, funding can be applied for from the Innovation Fund to bring the product to market maturity.

5.4 State research projects[37]

The Ministry of Innovation pursues a research policy that serves to develop urgent research questions. The people decide in a committee which research questions are urgent. State research projects are opened for subject areas that represent future challenges for the national economy and can be mastered through new technologies. Proposals can be made by research institutions and companies in the Research Directory as a profile and can be declared to be of interest to society as a whole. Which projects are selected is determined by the Minister of Innovation in voting with the people in a committee. The Minister of Innovation can also present his or her own research projects at the committee. Selected research projects receive state funding through the Ministry of Innovation's provision of research funds and the establishment of research communities[38] . Research funds for state research projects come from the Ministry of Innovation's own resources and from the Research Cost Fund.

---

37 §183.7 Research and innovation
38 Ministry of Education - 11.7.1 Research community

5.4.1 Provision for the future[39]

The research project for future precautions is primarily concerned with how a circular economy should be designed so that it is harmless to the environment and does not require finite raw materials. Within the framework of the project, research assignments are issued to scientifically investigate natural and artificial closed-loop systems. Human-made artificial closed-loop systems are being investigated for the most effective strategy.

The research fields further narrow down the search for circulatory systems. Research fields for the demands of the environment are climate, biodiversity, marine and coastal research, resources and geo research. Research fields for the demands of the population are the future of work and value creation, defence and security industry, civil security research for the prevention of danger, and the application of digitalisation in agriculture for the management of land with permaculture and indoor agribusiness.

The state's provision for the future is predominantly shaped by the need to provide the population with sufficient food, energy and shelter without generating hazardous waste that could make survival difficult or impossible for future generations.

The research task of the natural sciences is to investigate the circular economy in stable ecosystems in order to be able to transfer these techniques to man-made ecosystems. For example, natural habitats as a model and areas with permaculture and harvesting robots as a replica.

The research task of the social sciences is to explore the circular economy between production and disposal, but also between economic forms. For example, how the surpluses from the market economy eliminate the shortages in the Planned Economy.

The research task for the humanities is what role the individual should play in the circular economy, what its meaning is and how it should be understood.

The task of the Minister of Innovation is to clarify any contradictions that arise in a committee.

39§190.8 Environmental protection, §220.3g Agriculture: BV Art. 104

5.4.2 Digital transformation

The digital transformation is summarised in a research project that deals with the research and development of prototypes of digital systems. Research fields that can participate are digitalisation of private life and industry, security of digital systems, cooperation between humans and machines, artificial intelligence and autonomous machine behaviour. The natural sciences develop new technologies and test their impact on humans and the environment. The humanities develop healthy ways of dealing with and applying new digital technologies. The social sciences develop social and political manners and application possibilities, especially for digital communication technologies.

The Minister of Innovation is obliged to hold regular committees if the rights or health of humans could be endangered by new digital technology. Researchers and consumers are obliged to report discovered risks to the Innovation Agency.

5.4.3 Energy transformation

The research project for the energy transition aims to make natural energy sources usable by developing suitable energy generators and energy storage systems. Natural energy sources are the sun and water, which can also provide wind, fire and lightning.

All other natural energy sources, such as coal, uranium, petroleum, geothermal energy and natural gas, carry the risk of uncontrolled climatic change and health-threatening pollution, and they are also finite. Therefore, they are excluded from the research project and may be excluded from research throughout the country. Artificial energy sources are excluded, such as nuclear fusion, which, if used extensively, would have to extract vast amounts of deuterium and tritium and would release helium.

As long as the moon moves the seas and the sun warms the earth, there is enough energy available as long as it is harnessed through research and development. Energy generation through wave power plants, tidal power plants, current power plants, biogas plants, solar cells and solar power plants is promoted, as well as energy storage through pumped-storage power stations, zeolites and hydrogen. The state research project is limited to energy generators and energy storage systems that can be operated for as long as possible with as little digital technology as possible. Likewise, they should not continuously consume raw materials and continuously produce waste. In production, attention should be paid to renewable raw materials and recyclable materials. Biogas plants, which convert waste into electricity, heat and fertiliser, are exempt. With this research project, the Ministry of Innovation is pursuing the goal of enabling citizens to generate their own energy and to provide free energy for the economy.

In order to be able to dispose of the legacy of previous generations, research contracts are also awarded for the disposal of batteries, accumulators, nuclear waste and the dismantling of nuclear facilities.

5.4.4 Transport transformation

The research project for the transport turnaround explores human mobility behaviour and the future needs for transporting persons and goods on earth and in space. In cooperation with the driving and aircraft industry, research assignments are completed for new environmentally neutral propulsion technologies, refuelling options, land, sea, air and space vehicles, automated and connected driving and flying, digital land, sea, air and space management and the safety of these new technologies.

Existing research groups are the domestic Aerospace Centre and the European Space Agency (ESA)[40] .

5.4.5 Biology

The research project for life sciences includes research into new methods and technologies in the life sciences. Research tasks are to understand and be able to reproduce structures and

---

40http://www.esa.int/

processes of living organisms. The research results specified are material innovations that can be produced from renewable raw materials or that can grow to the desired form on their own in a nutrient solution thanks to a designed gene code.

The aim of this research project is to find renewable and biodegradable substitutes for steel and metal production as well as petrochemistry. Research fields are bioeconomics, bionics, biochemistry, biophysics and biotechnology.

5.4.6 Universe

The research project for the universe deals with the composition of the universe, the nature and interaction of matter and the resulting opportunities and risks for mankind. Research contracts are awarded for the exploration of elementary particles, for the disposal of space debris, the defence against celestial bodies and the search for a second Earth.

5.4.7 Basic research[41]

Basic research is conducted in the natural sciences, humanities and social sciences. The individual subject areas of the education and research institutions receive their orders for basic research from the Innovation Agency. It coordinates the distribution of tasks based on the qualifications of the participants involved through the community of domestic research institutions. For example, primary schools can participate in basic research on the plant world and conduct simple series of experiments with nationally specified plant species. The data collected is then statistically processed by comprehensive schools, analysed in colleges and published by institutes with possible links to applied research or industrial research.

Because basic research is very broad, citizens can also decide here in a committee which subject areas should be researched first. The Minister of Innovation ensures a balanced panel with 2 researchers from each of the natural sciences, humanities and social sciences, as well as an audience with 2 researchers

---

41 §183.6 Research and innovation

from each subject area. Other citizens can participate in the committee via state television. All citizens and companies can propose and rate subject areas in the Committee Directory before the committee, of which the 5 most popular ones on the committee will be turned into research projects. After the voting, it is decided which research projects will be the state's basic research.

State basic research is carried out in research institutions that are obliged to spend 10% of their annual research activity on basic research. This means that labour, premises, devices and materials are available, the costs of which are borne by the research institution because it uses the results of basic research. In order not to place an excessive burden on individual research institutions, special costs can be reimbursed by the Innovation Agency. Research institutions can also buy themselves out of the obligation to conduct basic research. This money is used to pay research institutions that carry out the lost basic research, as well as the special costs for research institutions that are overly burdened.

6 Innovation through education[42]

The ministries of innovation and education work closely together to promote innovation in and through educational institutions. Innovation in educational institutions means that the educational institution receives a research contract from a company in the Planned Economy or Social Market Economy or the Innovation Agency. The principal pays for the costs unless they are covered by the Research Cost Fund. Innovation by educational institution occurs when learners, teachers or researchers have an idea, research and develop it while working in the educational institution. Promising inventions may be pursued by educational institutions on their own authority and recruit investors and collaborators through a profile in the Research Directory.

The Innovation Agency is involved in the creation of the

42§176,1 Education space: BV Art. 61a, §177,2 School system, §180,2,3,6,7 Schools and colleges: BV Art. 63a, KV Art.44, §183,2,9 Research and innovation: BV Art. 64

curriculum to teach learners to be inventive at an early age[43] and to be able to cope with the future requirements in business and society instead of learning outdated techniques.

For quality assurance purposes, research assignments are completed as part of performance records and final theses so that a teacher checks the results. The Ministry of Innovation can advance a research project by having all graduates of a certain subject area nationwide complete orders for the state research project through their final theses. Quality assurance in the mobile Innovation Labs is carried out by the lab managers and the surveillance cameras. In research-based companies, the Company Auditing Agency ensures the quality of the research through its auditors for innovation and technology.

6.1 Inventions of interest to society as a whole

Patent examiners can determine the interest of society as a whole in an industrial property right or other invention. Inventors can declare the interest of society as a whole in their profile in the Ideas Directory, which the patent examiners or innovation auditors then check. This means that the invention could raise the standard of living of all mankind, if only it is implemented correctly. In the case of inventions of interest to society as a whole, a research assignment is sent out to all appropriate educational institutions. These orders are listed in the Education Directory and as soon as the keywords of the profile in the Ideas Directory match the subject areas visited by learners, the inventions are displayed to learners in their news as advertisements.

Learners self-select the invention of interest to society as a whole that they wish to research and develop. Their final results must be checked by teachers in performance records of the appropriate subject area. If the teacher responsible for the learner is overwhelmed, he or she can find suitable specialists through the Examinations Office to check the performance record.

The main question to be examined is whether the invention

43Ministry of Education - 9.15.1.2 Subject in the sixth learning year: Inventing is feasible and, if so, how exactly and, if not, why not. The teacher is responsible for contacting the other necessary subject areas at educational institutions to have the invention checked sufficiently scientifically. It may therefore take a long time before the return date, once the document has passed through the subject areas. Universities, institutes, colleges and comprehensive schools check, rate and optimise the technical feasibility so that the invention works or, if necessary, is classified as unfeasible. If in doubt, inventors can always contact the Innovation Agency.

6.2 Business cooperation[44]

The companies of all economic forms report their needs for research and development to the innovation auditors, who forward the information to the Innovation Agency. The field service checks which educational institutions are suitable to meet the needs and prepares cost estimates in voting with the Education Authority. The innovation auditors submit the cost estimates to the companies, which may then award a research contract or agree on a lease. The research assignment is carried out by the learners and teachers within the framework of the lessons and checked in performance records. The results are delivered to the company. The companies can also have their own employees conduct research in rented educational institutions, as long as no state research or training is taking place there. When capacity is fully utilised, only orders from companies in the Planned Economy and Social Market Economy are accepted.

6.3 Promoting innovation at educational institutions

The premises of the educational institutions can be used as experimental laboratories after classes. The People's Innovation Company Think Tank is responsible for the use of the premises. The People's Innovation Think Tank uses the Procurement Office and collaborations with companies to provide materials

---

44§181.3 Vocational training

and tools that are also used by industry. The aim is to make it easy for citizens and learners to participate in innovation development and to be inventive.

6.4 Invention studies[45]

The provisional patent entitles any inventor to study the necessary expertise at any available state college at any time in order to research his invention and make it ready for the market. Inventors are allowed to attend seminars in several subject areas of different state colleges at the same time.
If inventors know at which subject areas they can get the necessary expertise to find out if and how their idea becomes feasible, they can visit them. If they do not know, firstly there is the Knowledge Directory[46] , where all domestic educational programmes can be accessed digitally via the People's Computer. Secondly, there is a search function in the Invention Profile of the Ideas Directory. All educational institutions must keyword their educational programmes and seminar contents in the Education Directory and assign them to sectors. Via the invention profile, the inventor is automatically shown the matching events in his or her area with place, date and time. Users can register directly with the desired event via their People's Computer and are automatically linked to the appropriate group of the seminar in the Education Directory via their invention profile in the Ideas Directory and with their identity profile in the Persons Directory. Now the inventor gets admission to the detailed content of the seminar. If the content seems suitable to them, they can register for the seminar and attend it.
Student ID cards are issued after the responsible school administration has shown the invention profile of the provisional patent on the People's Computer. The matriculation number of this student is always also the number of the provisional patent.
The inventors are never obliged to take exams or to be

45§182.1 Continuing education: BV Art. 64a, §183.9 Research and innovation
46Ministry of Education - 12.7 Knowledge Directory

present because they do not intend to achieve an educational qualification. Instead, they have the right to ask the teachers present for help in researching literature or finding suitable experimental sites. If material is needed, an inventor has to pay for and obtain it himself. If he cannot pay for it, he can solicit money from the Ideas Directory or the Ideas Stock Exchange. If he cannot obtain material, he may order it through the subject area.

Inventors may also advertise to other learners and teachers. Teachers are encouraged to give students the opportunity to demonstrate their expertise through a rating of the invention and to assess this as an examination performance. For this purpose, the inventor must grant the examinee admission to his invention profile. If the student consents to the view, a non-disclosure agreement automatically applies. Students and teachers have the right to reject the offer. If the examination determines impracticability, the inventor is no longer entitled to continue studying at this subject area.

6.5 Mobile Innovation Labs[47]

Mobile Innovation Labs are designed to train voluntary researchers in as short a time as possible so that they can obtain sufficient expertise for the research and continue it at home. Mobile Innovation Labs are deployed in places where the relevant expertise is needed. The mobile Innovation Labs are equipped differently for this purpose.

In areas with few jobs or licensors, the mobile Innovation Labs are retooled for activities that can be carried out by the local population. To do this, data is retrieved on all the qualifications and interests of the affected citizens from their profiles in the directories for persons, education and labour.

The Ministry of Innovation maintains a fleet of 800 trucks whose interiors are adapted to the profits from the challenge search engine. For example, with the challenge: We need a battery that can be charged quickly and lasts a long time. In this case, at least one truck from the fleet will be equipped

---

[47] §182.1 Further education: BV Art. 64a, §212.5 Structural Policy, §183.9 Research and Innovation

with the various battery kits. A Truck tractor from the fleet will be equipped with camera cranes to broadcast experiment shows on state television.[48] The equipment of the laboratory can be designed very differently. It ranges, for example, from a biochemistry laboratory with all kinds of suction and isolation devices to a meeting room with a beamer, PC including statistical analysis programmes, a table and many chairs for social scientists.

Each individual Innovation Lab consists of two containers. One of them can be extended on both long sides and contains the laboratory. The second container contains the material storage and lockers for the inventors. Each lab is equipped with a 360° patent camera and microphone to prevent theft of ideas or materials. If a patent is granted, the film material becomes part of the patent application and facilitates the paperwork or helps to determine the number of inventors. There are also screens with instructions on the prevention of danger, i.e. which substances can be dangerous in the laboratory and when. For easy training and handling of substances, there are transparent training glasses[49] , which are as big as safety glasses. Warnings are automatically displayed.

There is one laboratory director and one laboratory assistant per laboratory. The lab manager is usually a doctor and the lab assistant is a research assistant, technical assistant or caretaker. Both are employees of a state educational institution who temporarily supervise mobile Innovation Labs.

The mobile Innovation Labs stand in squares near a town hall for 6 months, then they move on across the country. Unfinished workpieces can be locked up in the lockers. Those who have finished building things are allowed to register them as a provisional invention at the town hall and then take the thing home to market it or develop it further, for example at a local educational institution with similar laboratory conditions. Anyone who wants to order materials for their own further development can do so at their own expense via an order list, which must be on display at least once in every Innovation Lab, but can also be called up and ordered

---

48Ministry of Media Affairs - 13.2.4.3 Innovation Lab
49Ministry of Education - 12.4.1 Training glasses

in the Ideas Directory. Those who do not wish to apply for a provisional patent must leave their tinkering in the lab. The scientists on site decide which experiment should be included in their research or disposed of before reaching the next location.

7 Technology policy

The Ministry of Innovation operates a technology policy to test, certify, standardise and market new technologies in goods, services, workflows and digital systems. In cooperation with the Ministry of Education, it operates the Institute of Technology, which is responsible for developing and testing uniform standards.

The Company Auditing Agency's technical auditors test new products and check their safety, sustainability and quality in cooperation with the health auditors in the So-called conformity assessment.[50] Products that pass all the tests are given the PS (Proven Safety) mark if they are domestic and the CE mark if they are to be manufactured in the European Union or imported into the European Union.[51]

The innovation auditors certify new products and specify the release date of the novelty in the audit report. If they detect product piracy through unlicensed imitation or counterfeiting during the audit, they report this immediately to the legality auditors.[52] The innovation auditors identify obsolete technology and synergy effects during their audits and offer technology transfer via the Innovation Database. Technology transfer can be ordered through standardisation. For example, a company invents a catalytic converter for machines that keeps the air clean, then all similar machines must also be equipped with this new catalytic converter in order to obtain approval through a seal of approval. Standardisation is being driven forward above all in information and communication technology in order to be able to use digital contacts for data transfer with as many devices as possible. Safety standards are being set for internet architectures and digital product and

---

50 Ministry of Labour - 20.7.4 technical auditor, 20.7.2 health auditor
51 Ministry of Labour - 20.7.4.2 Seal of approval
52 Ministry of Labour - 20.7.6.4 Compliance with trade law

plant safety.

The patent examiners classify inventions in the appropriate industrial property rights and check whether all requirements of the affected industrial property right are met. For this purpose, the Patent Office determines the appropriate laws in voting with the Minister of Innovation.

7.1 Institute of Technology[53]

The Institute for Technology works with the Company Auditing Agency's technical auditors and innovation auditors, as well as with the Innovation Agency and the Institute for Evaluation[54] . The technical auditors also use their tests to collect data for large-scale studies conducted by the Institute of Technology. This allows products to be tested in practical use and provides complementary data to the laboratory testing of the product. The innovation auditors provide data on innovations in the state of the art and can market innovations from the Institute of Technology.

Every new product that is to be sold in series on the domestic market must be sent to the Institute of Technology for long-term studies. Measures for standardisation are introduced by the Institute of Technology to increase usability, sustainability and safety. Sustainability here means a long-lasting, repairable, environmentally friendly and healthy characteristic of the product. For example, plug shapes for chargers can be standardised. The Institute of Technology is responsible for norms within the country. It is audited according to the German industrial standard (DIN)[55] . For international norms, testing is carried out according to the requirements of the International Organisation for Standardisation (ISO) .[56]

Firstly, the Institute of Technology records the state of the art. The state of the art is investigated and documented in terms of material properties and processing methods. Secondly, it develops methods to be able to handle the technology properly.

53§183.2 Research and innovation: BV Art. 64
54Ministry of Labour - 20.10 Institute for Evaluation
55https://www.din.de/en
56https://www.iso.org/home.html

These methods include not only examination procedures for the technical auditors, but also instructions for use for the executives, the proper handling of which in companies can be confirmed by an examination by the Company Auditing Agency. And thirdly, it carries out long-term tests of all goods, services and work equipment on the basis of the ongoing tests of the technical auditors with this data. The data obtained in this way can be used to classify technology according to how long its useful life is at different frequencies of use. In this way, successful technologies can be found and their use can be adapted to the users.

The Institute evaluates the results of the Patent Offices and technical auditors. The evaluation is to examine successful new goods, services and means of work for their combinability.

7.2 Industrial property rights

The Ministry of Innovation determines what is considered an industrial property right and makes laws on copyright, trademark law, inventor law, design law and patent law. Industrial property rights can be acquired if an invention profile is created in the Ideas Directory. Employees of the Innovation Office can assist in this process. An industrial property right is applied for digitally at the Patent Office, which checks whether all requirements for the industrial property right are met. Industrial property rights are considered the property of the inventor until he or she sells it or does not maintain it. An amount must be paid to the Patent Office on a regular basis for maintenance, which differs depending on the industrial property right and the duration of maintenance. In the event of sale, the owner's personal details change in the Ideas Directory profile.

Industrial property rights are there to be used, not to prevent use. Preventing the use of an industrial property right can lead to procrastination of innovation and can be punished.[57] Other users must be able to acquire a licence. The licensor may set the price himself. If it is in the interest of society as a whole, the licence price may be set by the Patent Office.

57 Ministry of Justice - 8.8.3 Procrastination of innovation

7.2.1 Patent

A patent shall be deemed to be any technical invention which is a novelty, has been developed by an inventive step, is susceptible of industrial application and for which no other patent exists. An inventive step is anything which cannot obviously be assembled from existing material by a person skilled in the art. Novelty is guaranteed if the invention is not already state of the art and has not been displayed to the public in any way. The patent is valid for 20 years and must be renewed every 5 years. The renewal fee corresponds to a share of the profits of the past 5 years that have been made with the patent. After the first 5 years the fee is 10% of the profits, after 10 years 20% and after 15 years 30%.

7.2.2 Design

A design is any aesthetic invention whose shape, colour or form is a novelty and has a peculiar overall impression which distinguishes it from other designs even for informed users. The design is valid for 20 years and must be renewed every 5 years. The renewal fee is a percentage of the profits generated by the design over the previous 5 years. After the first 5 years the fee is 10% of the profits, after 10 years 20% and after 15 years 30%.

7.2.3 Copyright

Any literary, scientific or artistic work is considered a work of authorship. It is the property of the author for 70 years after first appearance until he is dead. Copyrights can be sold and inherited until the expiry of the term. A publication serves as a registration and the date of publication dates the beginning of protection. All subsequent identical works are considered counterfeits. A fee for registration or renewal does not have to be paid. Authors must mark their work with either ©, the year of publication and their name, which means that the commercial copying right lies with the author. The author must issue licences, but can set the price for them himself. Or authors who do not mark their work at all or mark it with (f), the year of publication and their name, thereby stipulating that everyone has the right to copy.

7.2.4 Brand

A trade mark is any word, figurative or word-pictorial mark. Word marks are brand names that can be written in any font. Figurative marks are graphic representations, also called logos or trademarks. Word and figurative marks are trade names in a separate font and graphic elements. An application for registration is not possible or is null and void if there is already an earlier identical trade mark with which confusion is possible. A trade mark can be protected for an unlimited period of time if the renewal fee is paid every 10 years. Upon request or failure to pay the fee, the industrial property right for the mark expires. The fee is 0.01% of the profits made with the trade mark, but not less than 100 euros.

7.2.5 Commercial process

Commercial processes are all processes that differ from other processes in their temporal, spatial and material application and are used for gainful employment. These can be, for example, special cooking recipes, computer games or formats of television shows. The industrial property right is valid for 15 years and must be renewed every 5 years after filing. After the first 5 years the renewal fee is 10% of the additional profits from the use of the commercial process, after 10 years 20%. The contents of the industrial property right must be digitised and sent to the Patent Office. There they are notarised and stored. Should unauthorised use occur, the similarity can be checked and an action for an injunction can be filed.

7.3 Patent Office[58]

The Patent Office conducts the granting and maintenance of industrial property rights and keeps statistics on them. The Patent Court is part of the Remit Court for Innovation and is responsible for all IP infringements. In voting with the European Ministries of Innovation, the Ministry of Innovation operates the Patent Office as part of the European Patent Office (EPO)[59] . The aim of European cooperation is to have uniform industrial property rights and uniform patent court procedures throughout Europe.

The Patent Office is responsible for examining all provisional industrial property rights submitted from the Ideas Directory as to their eligibility for protection, registering them and, after payment of the fee, ensuring that the industrial property right is maintained until the term of protection has expired.

If inventors are too often unable to select an industrial property right in the Ideas Directory and only indicate "other", the Patent Office examines the invention profiles that have indicated "other" as an industrial property right and can draft a new industrial property right in voting with the Minister of Innovation and put it to the people.

7.3.1 Promising patents

The Patent Office is the first official body to see a potential patent. As soon as examiners from the Patent Office have the impression that they are examining a globally groundbreaking invention, the examiners must submit the invention to the Minister of Innovation. Together with the Innovation Agency and the Minister of Labour, he decides whether the invention should become a new People's Innovation Company. If a majority of all the participants involved are in favour, the offer should be made to the inventor or inventors.

58§188.7 Statistics
59https://www.epo.org

7.3.2 Patent camera

When people want to talk about ideas to develop new products, they can go to the Innovation Office to lend out a 360° camera that can record videos. Wherever humans know in advance that they will share their ideas to develop an invention, the camera can be set up to record the contributions of all participants. This video should be saved in the invention profile if there is more than one inventor. The data from the camera is automatically sent to the Ideas Directory via the People's Computer and stored there. If there is no 360° camera for spontaneous inventions, all potential inventors film themselves with their People's Computer. Only the inventors and the Ministry of Innovation have access to these videos. As with all files, each access is stored in the Access Directory .[60]
Inventors can lend out the camera to document their invention. For this purpose, the Innovation Office also has a camera with a normal lens that inventors can use to produce a video for industrial property rights or marketing. There is a guide on how to present the product to be protected in order to describe it sufficiently for the invention profile. This guide is produced by the Patent Office and revised by the Ministry of Media Affairs to produce stylistically appealing films that can be tailored to target audiences. A video editing programme can be installed on the People's Computer and video editing programmes are available on the computers in the Intranet Café.

7.3.3 Register industrial property right

The registration of an industrial property right is always done on the intranet. Either on the People's Computer or in the Intranet Café in the town hall. To do this, the as yet unpublished invention profile in the Ideas Directory is adapted to the requirements for the corresponding industrial property right. The industrial property right can then be applied for. The Patent Office prepares a guideline for this purpose, which clarifies for each paragraph what should be written there and

---

60 Ministry of Digital Affairs - 7.5 Access Directory

how the illustrations are to be prepared. It also explains which features can be used to assign an industrial property right to a Locarno class, industry or company and how to assign the appropriate keywords. All this work takes place when entering profiles in the Ideas Directory and is done by the inventors or on their behalf by workers in the Innovation Office or patent attorneys.

As soon as this preliminary work has been done by inventors or suppliers, the Patent Office offers a comprehensive search with its searchers to find out whether the industrial property right already exists and to which protection class it belongs.

The Patent Office then offers advice by telephone with its in-house patent attorneys, during which the invention profile in the Ideas Directory is reviewed and orally corrected by the patent attorney of the appropriate field. The inventors or service providers receive the oral corrections in writing through a speech recognition programme in the appropriate field in the invention profile of the Ideas Directory, but must revise the profile themselves. Afterwards, they can call the telephone advice centre again for the final check to ensure that the description of the industrial property right is formally error-free.

In the final step of the application, Patent Office examiners check the invention profile for consistency of content, i.e. whether all criteria are met to develop an industrial property right from the invention. If this is not the case, defective or deficient parts are marked in the invention profile. Inventors or their service providers can now try to remedy the defects and have the industrial property right newly examined. Auditors may refuse to grant the industrial property right after the second examination and mark it as "not eligible for protection". These cases are forwarded to the New IP Rights Department at the Patent Office. There it is decided whether a new industrial property right must be created or whether an invention is marked as "null and void". An industrial property right is null and void if the criteria for no existing industrial property right are met and the invention is not marketable. To check marketability, provisional patent protection can be granted and the invention is published on the crowdfunding platform of the Ideas Directory. If there is sufficient popularity of 0.1% of the population, an approximate industrial property right is granted or a committee is convened. Any designation as "null and void" can be appealed to the patent court. Once an industrial property right has been marked "granted" by auditors, the invention profile can no longer be processed.

7.3.4 Inspiration Fee

If a product is similar to a protected invention, the inspiration fee must be paid to the owner of that industrial property right. This does not have to be done if the industrial property right has expired. The inventor is already automatically informed of this when filing the application. The amount of the inspiration fee corresponds to the renewal fee of the industrial property right after the first 5 years. The inspiration fee is payable only once and can be paid within 2 years. The similarity must be between 50 % and 90 %. All similarities below 50 % do not need an inspiration fee, provided that no essential features have been copied. Similarities above 90 % are not possible, but then a licence must be obtained from the owner of the similar industrial property right.

8 Ideas Directory

The Ideas Directory is a mixture of patent databases[61], Kickstarter[62] , Amazon[63] and Facebook[64] . Inventions are given their own profiles. Likewise, all valid and expired industrial property rights are given a profile. For this purpose, all databases of the Patent Offices worldwide are regularly accessed, the content of each entry is copied and inserted into a new profile in the Ideas Directory. This shows the state of the art, when industrial property rights expire and are then in the public domain.

Inventors can create a new profile with their People's Computer

---

61 https://www.epo.org/searching-for-patents/technical/espacenet_de.html
62 https://www.kickstarter.com/
63 https://www.amazon.de/
64 https://de-de.facebook.com/Directory/places/

or in the Intranet Café. If possible, they should do this as soon as they have the first idea. The profile is intended to accompany inventors from the first idea through the appropriate industrial property right to marketing.

8.1 Inventor groups

In the Ideas Directory, inventors can jointly create their invention profile and must form a group for this purpose. All inventors who are involved in the idea generation process must agree to join the inventor group with their People's Computer. The group itself determines after each meeting how high the contribution of each group member was, i.e. the inventiveness level of the ideas that were contributed. In all groups, confidentiality agreements apply in principle, so that no group member may in any way inform outsiders about the invention.

8.2 Invention profile

The invention profile is in different stages of development. A bar above the profile shows which stage the invention is in.

8.2.1 Profile visibility

Until registration, the profile is only visible to the inventor. If there are several inventors, they form an inventor group that can jointly edit the profile. The visibility of the profile can be released at any time, but this jeopardises the novelty required for industrial property rights. In the data security settings, visibility can be released for groups of people. These are the categories: Only me, Only the inventor group, My friends in the Persons Directory, My business contacts in the Ideas Directory, All in the Labour Directory, All in the Persons Directory.
If an industrial property right is to be applied for, only inventors and employees of the Ministry of Innovation can view the profile and edit it for a limited period of time within the scope

of their duties. Edits by staff of the Innovation Agency or the Patent Office must be displayed to the inventors in the access log[65] . Once an industrial property right has been granted, the profile can no longer be edited and is visible to all users of the Ideas Directory. Only the name of the owner and licensee can be changed subsequently if the industrial property right is sold or marketed under licence.

8.2.2 Logging

The People's Computer can be used to record the brainstorming process. The protocol can be created by a text, a drawing, a sound recording, a photo or video. Videos are obligatory when logging in an inventor group. A link in the invention profile starts the video and uploads it in real time to the Ideas Directory server. This is the only way that no post-processing of the video can take place and the video can be considered a court-proof record. Users decide whether to use the People's Computer camera or the patent camera from the Innovation Office, depending on the situation during the idea generation process. When one finishes recording, the video file is saved on the Patent Office server and additionally on the People's Computer. Those who do not use a People's Computer can lend out a patent camera and a universal People's Computer from the Intranet Café. For use, the identity card must also be expelled into the card reader slot here.

This step is primarily for record keeping and is non-binding because inventors can add or discard files at any time. Advice and equipment can be obtained in person at the Innovation Office or digitally spontaneously with any People's Computer via admission to the Knowledge Directory[66] .

---

65 Ministry of Digital Affairs - 7.5 Access Directory
66 Ministry of Education - 12.7 Knowledge Directory

8.2.3 Categorisation

The categorisation consists of a questionnaire about the invention. In it, various choices for economic or social areas, such as goods, services, industries or remits are offered for selection. Through the automatic search, all previous information in the profile, including the protocols from the idea generation, can be used to be displayed as an answer option in the questionnaire. The questionnaire asks for all the requirements that make up the various industrial property rights. In the Innovation Office and in the Ideas Directory, there are completion aids for the questionnaire. Any unclear additions can be made after reading the guide to describing an invention or viewing the training videos on your own People's Computer. At the end of the questionnaire, all industrial property rights that may apply to the invention are listed.

8.2.4 Selection of the industrial property right

The inventors indicate in the invention profile in which areas they would like to have protection for an invention. They can choose between patent, design, copyright, trade mark, industrial process or other. During the examination at the Patent Office, the inventor is offered all possible protective industrial property rights, but only the industrial property right indicated is examined. If you do not know which industrial property right is the right one for your invention, select "Other". The selected industrial property right is displayed at the top of the status bar. It is possible to acquire several industrial property rights for one invention. For example, the name of the musical instrument "Hang" is protected as a brand name, the shape is protected as a design, the copyright work enjoys copyright protection and the production process is patented. In this case, the invention profile would have four industrial property rights.

8.2.5 Description

The description of an invention must be uniform for each industrial property right in order to facilitate the subsequent search in the database and categorisation. The inventors learn which requirements apply to the selected industrial property right from a compilation of learning videos, from viewing other invention profiles with the same industrial property right and from a filling-in aid. The filling-in aid asks the inventor the appropriate question for each area. The inventor must write or speak the answer in the input field, whereby a voice recognition system immediately transcribes the spoken words. If possible, drawings, photos or recordings of sounds or videos of the invention should be attached and described in the description.

8.2.6 Own investigation

You can carry out your own investigation manually at any time by entering search terms and categories and browsing search results. To complete your own investigation, you must first perform a manual search and then an automatic search. The automated investigation can only be carried out once the categorisation, the selection of the industrial property right and the description have been completed. The algorithm accesses all these details, categories, image and sound files. Drawings and photos are recognised and matched by the algorithm of an inverse image search function[67] . If the inventor has investigated well, the automatic search will show him results that he has already seen. If the inventor has formulated poorly in the description, he will be shown unsuitable inventions. The automatic search classifies the search results into "similar" and "too similar". Search results that are too similar make the invention ineligible for protection because it is no longer new or already protected. All search results that the inventor or the automatic search classifies as too similar can be saved in the profile. If the inventor is unsure whether the other protected inventions are too similar to his, he can call the fee-based

---

67https://www.google.de/imghp?hl=de&tab=wi&ogbl

search number, where Patent Office searchers compare the profiles marked as "too similar" with the inventor's profile. If they confirm that the profile is too similar, it must be deleted or cannot be protected. The profile only has to be deleted if there is a valid industrial property right for the invention that is too similar. The invention cannot be protected if the industrial property right for the invention that is too similar has already expired and the invention is therefore in the public domain.

8.2.7 Legal examination

In the legal examination, patent attorneys from the Innovation Agency sit in a telephone centre and check invention profiles for formal faults that could become contentious in the event of court proceedings. The conversation begins in the secretariat, where the number of the profile is requested. This number enables a programme to enter the invention profile in the Ideas Directory and match the categories and keywords given with the lawyers' respective specialist departments. The lawyer now has the documents directly in front of him and reads them through while the inventor is on the phone. During this process, the lawyer's eyes are filmed by a camera, which directly shows the inventor where the lawyer's eyes are on the patent document at that moment. As a program function, the lawyer has the possibility to mark passages that are incorrect with a virtual red pencil. In this case, the lawyer pronounces a correct wording, which is recorded by a speech recognition programme and displayed as text on the screen of the People's Computer and can subsequently be entered into the profile by the inventor(s). For this telephone call, a price per minute applies that is equal to the lawyer's hourly wage divided by 60, plus 10% profits for the Ministry of Innovation.

8.2.8 Investigation by the Patent Office

Industrial property rights can only be granted on unique things. To check for uniqueness, Patent Office auditors conduct an investigation in all international industrial property right databases. Once they are done with the search process, they save the invention profile and note in the profile whether the invention is unique or not. If it is not unique but already protected, the searchers send a message to the inventor. In this message, they request the author of the profile to delete his profile immediately and refer to the similarly valid industrial property rights found.

The Patent Office is responsible for making as many innovations protectable as possible so that inventors can be remunerated for them. If "Other" is selected as the industrial property right, the searchers assign all suitable industrial property rights to the invention. The inventor must then decide which industrial property rights to apply for. If there is an innovation but no suitable industrial property right, the searchers report this to the Patent Office management so that a new industrial property right can be developed if such innovations occur more frequently. The costs for the investigation amount to the cost price for the searchers plus 10% profits and are due as a fee when the order is received.

8.2.9 Examination

After the investigation has been successfully completed and the inventor knows which industrial property right he wants to apply for, he can order the examination of the invention for the specified industrial property right at the Patent Office. The inventor can indicate whether the industrial property right is to be applied for in Europe only or in the whole world. The Patent Office's auditors will check whether the invention is protectable and whether all the requirements have been met. If certain changes are necessary for an industrial property right to be granted, the examiners report this to the inventor by marking and commenting on the passages in the invention profile. The inventor can make these changes, but must then commission a newly examination. The cost of the examination is the cost price for the examiners plus 10% profits and is payable as a fee when the order is received. After the examination, the invention profile is deemed to be "protected" or "not protectable" and the industrial property right is granted or not.

If the industrial property right is to be filed worldwide, the Patent Office will file the application worldwide and also provide the necessary translations. This service is subject to a fee and corresponds to the cost price of the Patent Office plus 10% profit.

8.2.10 People's Innovation Company suitability

When examining patents, the patent examiners always also check whether the patented invention is so innovative, socially desirable and marketable that it should be used to establish a People's Innovation Company. Therefore, when a patent is granted, the indication "People's Innovation Company examination positive" or "People's Innovation Company examination negative" is also given. If the People's Innovation Company examination is positive and the inventor has already paid for the worldwide application, these fees will be refunded if he accepts the offer and has the patent marketed by a People's Innovation Company.

8.2.11 Publication

Inventors who wish to make their invention freely accessible, i.e. waive an industrial property right, can publish their invention profile at any time. As soon as an industrial property right has been granted, the invention profile is published in any case. Inventors can decide whether only the necessary information for an industrial property right is published or also other information, such as protocols and sketches. In the case of a patent, publication can be deferred for up to 12 months. During this time, the patent enjoys worldwide protection, the So-called priority. All granted industrial property rights are also published in the Patent Office's internet-based database. For this purpose, all necessary information from the invention profile is transferred from the Ideas Directory to the database. As soon as a new patent is published in the Ideas Directory, all People's Computer users can rate the patent. The 100 patents that had the most positive ratings after 8 months are filed for international patent at the state's expense.

All visitors to the Ideas Directory can enter a price they would be willing to pay for an invention on the invention profile. Those who enter their price will be informed of the median of all price entries. All contributions, ratings and comments from users are published on the invention profile's pinboard.

8.3 Crowdfunding platform

Investors and employees can be recruited on the crowdfunding platform. As with startnext[68] or kickstarter, an invention is presented in a video and money can be donated in different amounts. Investors can be partners who help run the company, supporters whose monetary contribution is rewarded in kind as a thank you, or creditors of bonds for the company or the product in which the invention resides. Employees can be skilled workers, providers or buyers. Depending on what inventors are looking for, the relevant persons can report to the inventor via the invention profile.

8.4 Challenge search engine

The challenge search engine is an ongoing voting on the most pressing challenges that can be solved through research and development. Users can describe a current challenge for society in a maximum of 50 characters and assign up to 10 keywords. Each challenge must be sorted into at least one economic sector or scientific subject area. Similar challenges are grouped automatically. Authors of similar challenges will be shown the previous description and can join it. However, if their challenge differs from the one displayed, it can be

---

68https://www.startnext.com/

published. Other users can confirm or reject these challenges. Matching inventions from the Ideas Directory and research projects from the Research Directory are displayed for all challenges. Users can rate whether they think the invention or research project is appropriate for the challenge. As soon as 30% of those entitled to vote support a challenge, a committee is held on the necessary research projects or innovation funding. As soon as 50% of those entitled to vote consider an invention or research project to meet a challenge, a vote is held on whether the invention should be produced in a People's Innovation Company or made a law, or whether a research project should be implemented in a nationwide research community and funded from the state treasury.

The data from the challenge search engine also benefits companies, inventors and those equipping the mobile Innovation Labs. Companies can produce suitable products and win customers. Inventors can adapt their inventive activity to the challenges. Mobile Innovation Lab outfitters can facilitate the exploration of popular challenges with matching materials.

9 Innovation promotion[69]

The Ministry of Innovation promotes innovation by supporting the creation of innovative companies, the introduction of innovations into existing processes and the transformation of outdated processes into advanced ones. Support is provided through relevant research and development, advice, networks, industrial communities and funding opportunities to introduce innovations. To bring innovation support to inventors, companies and individuals, the Innovation Agency works with them. The field service ensures cooperation between education and research institutions and companies conducting industrial research. The Research Directory and the Ideas Directory provide the necessary innovation networks. Sufficient innovation advice is provided in the Innovation Office and through the Patent Office. The innovative industry policy spans all economic forms and uses them in a targeted manner to make innovation successful in the market. The

69§183.1 Research and innovation: BV Art. 64

indoor service of the Innovation Agency is responsible for voting on the measures with the ministries of labour, finance and economy. The capital goods industry, in particular, can use the funding opportunities in the Ideas Directory, Ideas Stock Exchange and Innovation Database to market equipment that reduces costs and eliminates damage to humans and nature. Innovation promotion also includes marketing through advertising in the Ideas Directory, in shops, at trade fairs and through award ceremonies.

9.1 Marketing inventions

Once the appropriate industrial property right has been registered, the invention can be marketed via the Ideas Directory and the Labour Directory. Either inventors can try to win companies as buyers, licensees or producers, or companies can ask inventors if they can produce the invention under licence or for the inventor.

9.1.1 Companies network with inventors

Companies can subscribe to news about new inventions relevant to their industry or product range in the Labour Directory. As soon as industrial property rights are filed or revoked in the selected areas, a notification is sent with the link to the corresponding invention profile in the Ideas Directory. If companies are interested in an invention, they can contact the inventor directly via the invention profile to produce the invention for them or order a licence.

9.1.2 Inventors network with companies

Inventors can use the invention profile data to conduct a search query in the Labour Directory to find companies that could use the invention under licence or produce it on orders. As a search result, all companies found are displayed in a list that can be sorted according to the turnover of the company, similarity of the product range, satisfied customers and the radius. Industrial property right holders can send an automatic request to all selected companies.

9.1.3 Self-marketing

Inventors can always set up an Innovation Enterprise in a Social Village if they are unsure about setting up and running companies.[70] If they wish to market the invention themselves, they must set up a company in the economic form of their election and create a corresponding profile in the Labour Directory. Necessary information is automatically retrieved from the invention profile in the Ideas Directory. The search functions in the Labour Directory can be used to find workers, machines, commercial properties, providers or buyers. Direct marketing is possible via the webshop functions[71] , which enables sales to users of the intranet and internet.

9.2 Innovation news

The Innovation News is a computer programme available on the intranet site of the Ministry of Innovation in the State Directory. It allows you to search for researchers, inventors, entrepreneurs, research projects, inventions and entries in the Innovation Database.
The Innovation News contains all newly filed industrial property rights from the Ideas Directory and all completed research projects from the Research Directory, as well as all companies from the Labour Directory that are looking for such innovations.
If there is no intranet access, a notification order can be deposited in the Innovation Office of the town hall by post or email that news has been received for innovations sought. However, this service is subject to a charge, unlike the digital companion application.

---

70 Ministry of Planned Economy - 10.6 Innovation Enterprise
71 Ministry of Labour - 13.2.2 Digital marketing

9.2.1 Subscription

Via a subscription, the search queries can be saved as a basic interest. Regular updates then follow. An update means that the search query is newly performed and shows all innovations since the last update. Users are asked to rate profiles from the directories for ideas and research so that the most popular innovations can be broadcast in the business section of the news .[72]

9.2.2 Search settings

Researchers, inventors and entrepreneurs can describe their interests and search areas in keywords in the Innovation News settings to help the algorithm deliver better results. Innovation News is available as a companion application for the Labour Directory, Ideas Directory and Research Directory and enables comprehensive search and targeted information.
In the search settings, companies, researchers and inventors can also specify when industrial property rights that interest them expire. They can set how long before the expiry date they would like to be informed in order to prepare in advance, if necessary, for production, research projects or further inventions that would have infringed the industrial property right beforehand.

9.2.3 News for companies

Companies can order innovation news tailored to them via the Labour Directory. Companies can specify what they are looking for in keywords. This search engine accesses the keywords and all the data in the company profile and matches them to existing categories and keywords in the invention profiles.
If a sought-after invention is to be marketed under licence, this report is published immediately and sent to all interested companies that have selected in their Labour Directory profile the areas in which they require innovation.

72Ministry of Media - 8 News Television

9.2.4 News for inventors

Inventors receive their innovation news via the Ideas Directory when they call up their invention profile. The keywords from the invention profile are linked to the sector in which matching companies are active. This makes it easier for inventors to find companies, licensees or producers. After the initial overview of all companies, inventors receive ongoing news about newly founded or closed companies in their field of activity. Inventors also receive news when a new invention is similar to theirs.

9.2.5 News for researchers

Researchers receive their innovation messages via the Research Directory when they open their research projects. Matching news is displayed by evaluating keywords from the Research, Ideas, Education and Work directories. This makes it easier for researchers to find workers they can work with, companies that can become buyers of their research results, or research institutions and research projects that are looking for suitable researchers. After the initial overview of all matching companies, research institutions, inventions and research projects, researchers receive ongoing news about newly founded or closed companies in their field of activity.

9.2.6 Keyword cloud

In order to give researchers, entrepreneurs and inventors an overview of the current innovation market, keywords of research projects, invention profiles and expressions of interest from companies are displayed in a keyword cloud. The keyword cloud can be set in four different levels, namely for research projects, inventions, company applications and all three previous levels together. In each individual level, words are displayed in different sizes. The largest words were mentioned the most, the smallest the least. In the fourth level, all matching research projects, invention profiles and company applications are displayed in red.

Researchers and inventors thus recognise sought-after ideas and other researchers and inventors who are advancing research and developments in the same field. Both can then join together in research groups or inventor groups.

Companies can see if many of their competitors or partners are looking for the same idea. This way, companies that are looking for the same thing can join together and pay into the Innovation Fund or the Research Cost Fund, or set up an Innovation Community. As soon as there is a new invention that companies are looking for, they receive a message with a link to the invention profile. The same applies to research projects.

9.3 Innovation workshops

If a production site and staff are needed for series production, but no money is available for this, inventors can set up an Innovation Enterprise[73] and use the Innovation Workshops[74] of the Social Villages as their first production site. Companies can use the innovation workshops for a fee to produce a pre-production mainly with their own staff. If they need additional staff, they have the same rights and obligations as inventors to recruit Social Villagers as employees.

The inventors are allowed to build their inventions in the innovation workshops and employ the residents as employees. As soon as the company generates revenue, they pay a wage that must be higher for each employee than for inventors. This work is offered in the work area luxury supply[75] . If the product or service makes it to market, the inventor and the staff use the profits to buy or rent a production facility outside the Social Village. This means that working in the innovation workshop makes all participating Social Villagers potential employees in the newly emerging company.

Each Social Village has different special tools in its innovation workshop, which can also be exchanged between the Social

---

73 Ministry of Planned Economy - 10.6 Innovation Enterprise
74 Ministry of Planned Economy - 10.9.4 Innovation workshops and laboratories
75 Ministry of Planned Economy - 10 Work area luxury supply

Villages. For production, the machines are to be moved or, if necessary, the employees. Each innovation workshop always has a 3D printer and a CNC milling machine together with the necessary computers to support these devices with data. Inventors and owners of industrial property rights automatically have admission to all innovation workshops of all Social Villages for 12 months and can report the need for employees at the gate. This is followed by a report in the Social Directory and a posting of this job vacancy on the Social Village notice board, the broadcast of a commercial on Social Village Radio six times a day for a week, and the announcement at the weekly plenary assembly of the Social Village on the following Sunday.

The right to use the innovation workshops expires as soon as the turnover is sufficient to run the business outside the Social Village. In which of the other three economic forms the company is to be opened and who is to work there is decided democratically by all those employees who have jointly earned the profits for the market entry costs.

Companies that have made a pre-production leave the innovation workshops as soon as the profits from the pre-production are sufficient to build up production at the company location. Social Villagers must be offered employment with the company if new workers need to be hired to produce the innovation at the company site.

9.4 Promotional video

Inventors who have an invention profile can have a 30-second promotional video produced for them by the Ministry of Media Affairs.[76] The promotional video will run once free of charge after the business news on News Television[77] . The cost of the promotional video consists of the labour and material costs for production plus 10% profits for the Ministry of Media Affairs. After completion, the promotional video is saved on the invention profile and can be downloaded from there and shared on the crowdfunding platform or in other

---

76 Ministry of Media Affairs - 5.3 Advertising
77 Ministry of Media - 8 News Television

directories.

9.5 Screen advertising

The screen advertising is offered in shopping shops. Shopping shops of the Social Market Economy have to put up the So-called innovation screen in their shop. Depending on which assortment the shop offers, only the invention profiles with the corresponding product groups are displayed. If a promotional video is available, it will also be shown. If possible, the innovation screen should be placed where the most similar products are placed in the shop. Marketing via screen advertising is similar to the advertising methods used in DIY stores, where videos of products on display are shown. With the innovation screen, however, the product shown is usually not displayed in the shop. It can be rated and ordered via the screen.

In the rating, users can also indicate a price at which they would be willing to buy the innovative product. The collected data is made available to the person who wants to market the invention. This can be inventors or companies that take out licences.

The price for orders is made up of production costs and profits for inventors, producers, shop owners and 10% of the price for the Ministry of Innovation. Orders are processed via the web shop from the Labour Directory company profile, but can either be paid for there and delivered to the worker's home or paid for in the shop and picked up.

9.6 Domestic fair location

The entire country serves as a huge exhibition space for new technologies. All technology developed domestically is also to be used here, at least once. Every tourist is thus also a trade fair visitor. The country is the presentation area, the production and sales halls are the stands, the hostesses are the citizens and the products are the property of residents or resident companies. The invention profile indicates where an invention is used and can be observed in use.

The Minister of Innovation conducts a committee when deciding which domestic innovations should be exhibited at the World Exhibitions (EXPO). In the committee, the most popular inventions are selected according to the election of persons process[78] , except that here the candidates are replaced by inventions and their intended use replaces the programme. The Ministry of Innovation holds an innovation fair once a year. Participation by visiting or having their own stand is free of charge for all exhibitors who have acquired an industrial property right in the past year or who wish to offer their invention free of charge. The exhibitors are responsible for equipping the stand with personnel and materials as well as for setting up and dismantling. The exhibition space is distributed equally, exceptions are possible for large inventions.

Otherwise, the mobile exhibition stand stands in the capital city of the Ministry of Innovation all year round and houses all inventions created between two World Exhibitions, at least as a virtual model. Afterwards, they go to the Museum of Innovation.

9.7 Innovation Database

The Innovation Database is a link from all innovation databases of individual companies. Each company can create an Innovation Database via the profile in the Labour Directory. Via the Ideas Directory, all Innovation Databases of all companies from the Labour Directory are combined as one Innovation Database. This database is stored on the server of the Ministry of Innovation and offers unlimited storage space.

9.7.1 Contents

This database is used to digitally store company developments and purchased industrial property rights or licences. In-house developments are innovations for which no industrial property

---

78 Ministry of State Organisation - 9.9 Elections of persons

right can or should be registered. These include trade secrets, innovative processes for production, services, marketing or accounting, as well as all inventions and copyrights in the making. All content stored in the Innovation Database is notarised. A time and date stamp is added during the storage process so that it is clear who had an innovative development first.

9.7.2 Entries

Entries can be made in the Innovation Database by entrepreneurs and innovation auditors. Between audits, entrepreneurs and employees can continuously document and save company developments. All entries since the last audit are examined by the innovation auditor during the audit, rated and notarised in the audit report. Thereafter, proprietary developments are considered innovations. Innovations discovered during the audit are certified and registered directly by the innovation auditor.

9.7.3 Scopes

Each innovation receives a profile in the database, which is divided into a secret and public area. In the secret area, the new product, service or workflow is described, shown and explained. The public section in the profile may be shown to innovation auditors and is published in the Ideas Directory and on the company profile in the Labour Directory. There it is shown for which purposes the innovation can be used and whether or at what price a licence is available.

9.7.4 Access rights

Only the innovation auditors have access to all content, including all secret areas. Each company has access to its secret area, the content of which can be marketed by the innovation auditors with the company's permission. All public areas are accessible to all users of the Ideas Directory.

9.7.5 Licensing

Innovations can be shared with other companies through licences. Entrepreneurs can determine with which economic forms, industries or companies the innovation may be shared, at what price and for how long. If the proprietary development was invented by employees, at least 20% of the licensing revenue must be paid as employee inventor compensation to the employee(s) concerned. The ministries of economy issue regularisations on how many or which innovations companies share with each other immediately, on request, not at all, for free or for a fee.

9.7.6 Seller

The innovation auditors ask the entrepreneurs whether they are willing to share their innovation under licence with other companies. If the entrepreneurs are willing, the innovation auditors sell these innovations in other companies as part of their audits.

If entrepreneurs want to buy an innovation after their own investigation in the Innovation Database, they contact the innovation auditors. They then submit an offer to the entrepreneurs after they have viewed the secret part of the innovation profile. The offer also includes an automated assessment of the prospects of success for the interested company.

After each audit with shortcomings, the innovation auditors offer suitable innovations from the Innovation Database, Ideas Directory and Success Model Directory that could benefit this company. They ask whether a company with shortcomings would like to buy another company's solution. If the answer is yes, the innovation auditor asks whether it is allowed to sell to this company. If the answer is yes, the innovation auditors may disclose the development and no longer violate their duty of confidentiality. Now the defective company can also reject the offer, but will then be specially examined at the next examination to see whether it has illegally copied the offer.

The sellers of licences are the innovation auditors of the

Company Auditing Agency. A share of 10% of the licence fees is due as a brokerage fee, of which the innovation auditors must pay 99% to the Ministry of Labour and receive 1% as a brokerage commission.

9.7.7 Buyer

Since the companies may also be offered unpublished trade secrets from the Innovation Database as a licence, and have thereby learned them in whole or in part, they are under a duty of confidentiality. If the entrepreneurs refuse such an offer, the innovation auditors check in subsequent audits whether the companies have made a pirated copy.
If the purchase is made, the innovation auditors help to introduce the innovation in the company taking the licence. At best, the innovation auditors who once notarised the innovation in the licensing company are used for this.
The innovative processes should also be treated as trade secrets in purchasing companies so that non-paying domestic and foreign companies cannot copy the innovations.

9.7.8 Prices

The price of a licence price table in which the developer places his innovation applies, plus 10% profits to the state treasury. These prices depend on the time of use and the increase in turnover of a licence and are payable once or annually.
The aim is to keep the prices for innovations lower than the resulting profits. This creates incentives to become more innovative and make more profits.
Companies in the Planned Economy are obliged to share free of charge with each other, companies in the Social Market Economy have a fixed price table for sharing and companies in the Free Market Economy have freedom of contract.[79]

---

79 Ministry of Free Market Economy

9.8 Innovation meetings

Innovation meetings are organised by companies and ministries and are designed to solicit ideas from outsiders. All companies can hold innovation meetings, in Planned Economy they have to do so. Every citizen is allowed to suggest improvements to companies. If you are unsure about which company an improvement proposal fits or where and when innovation meetings are held, citizens can find out at the Innovation Office.

The innovation consultation hour is possible in two forms. In the first form, an idea generator comes to the consulting hour and meets a staff member who is authorised to remove the responsible technical personnel at short notice. The two or more persons consult in the presence of at least one People's Computer to film the conversation and upload it to the Ministry of Innovation's server.

In the second form, all ideas received on an ongoing basis are collected and once a year a workshop lasting several days is organised. All those who have ideas, whether outsiders or employees, are invited, as well as all employees who are involved with the idea in their work processes. The aim is to make the ideas useful for the company.

The Company Auditing Agency's innovation auditors examine all the innovation consultation data collected by the company in the upcoming audit and collect their own data via the questionnaires to employees and customers. All contributions that can be used as innovations are notarised and entered into the company's Innovation Database. Idea contributors receive feedback on whether and how their idea was contributed. The inventors of an innovation are entitled to have it marketed by the innovation auditors via the Innovation Database, as long as the company that organised the innovation meeting does not conclude an exclusive contract with the inventor.

If a company already possesses a proposed idea, the existing idea must be proven to the innovation auditor as an entry in the Innovation Database or in the operating procedure.

9.9 Innovative employees

There are humans who perform work while constantly improving their work steps. These humans are called innovative employees. They are to be found through the Company Auditing Agency's employee questionnaires in all economic forms. The innovation auditors check the innovation skills of the employees during the regular audits. In the presence of the entrepreneur, the employees are to describe how the work steps looked before and how they look after his innovations. If the achievements can be confirmed, they are mentioned in the audit report as innovative employees and receive invitations to the innovation congresses.

9.9.1 Innovation congress

At innovation congresses, all innovative employees who were or are in the same or similar workplace meet and develop the most efficient work process at the congress and adapt it for their company. All companies with these work processes then benefit. The innovative employees get overtime for the congress and can achieve higher wages in wage negotiations through the employee-inventor compensation. At each innovation congress, the Company Auditing Agency has a stand where innovation auditors advise the innovative employees on how best to market their ideas.

9.10 Cartels for innovation

The Innovation Agency, in cooperation with the Antitrust Agency[80], ensures that companies can join together in Innovation Communities and form a cartel for a maximum of 20 years to market the innovation. Price agreements, purchasing cooperations, non-compete agreements, exclusive purchasing or supply obligations or market divisions are then not illegal, but must be approved and monitored by the Antitrust Agency. In the Barter Economy[81] and Planned Economy they are even

---

80 Ministry of Labour - 15 Antitrust Agency
81 Ministry of Barter Economy

necessary to organise work and avoid exploitation. In the Social Market Economy they are only allowed temporarily until an innovation has been successfully imported into the market. The innovation auditors check whether a temporary relaxation of the antitrust law[82] is permissible and instruct the economic auditors.[83] The Company Auditing Agency's economic auditors then check the total investment costs that were necessary to bring the innovation to market. Once 120% of the investment costs have been recovered, the Social Market Economy's antitrust law takes effect in the usual way. In the Free Market Economy, cartels are completely prohibited, unless companies of the Free Market Economy cooperate in a People's Innovation Company.

The Ministers of Economy of the Barter Economy, Planned Economy and Social Market Economy have the right to reduce the prices of cartels, but not below the limit of a profit zone of 5% of total turnover. The business consultants of the Company Auditing Agency assist the ministers in the decision-making process. The companies have to negotiate pricing with the citizens in a committee when there is a quorum of 50% of the affected citizens.

9.10.1 Innovation Community (IC)

The Antitrust Agency approves the merger of companies within one or more economic forms for the purpose of joint research, development and marketing of an innovation.

Companies wishing to set up an IC apply to the Company Auditing Agency's innovation auditors with their product and potential product range, which, if approved, will entitle the Antitrust Agency to set up an IC. They have to check how expensive and risky the development will be or how necessary the unification of patent rights is to connect technologies that make the novel product possible in the first place. The innovation auditors always ask themselves whether this product will provide mankind with technical progress that is

---

[82]Ministry of Labour - 20.7.5.6 Approval of relaxations of anti-trust laws
[83]Ministry of Labour - 20.7.3 Economic auditor

necessary to raise their standard of living. Another condition is that production is only carried out domestically by domestic companies and nationals. This is necessary to protect the admission to the innovations that are only available through the People's Computer. In this way, data security and company secrets can be protected from international access outside domestic jurisprudence.

ICs are also approved whenever the necessary majority for a new People's Innovation Company to be established was not achieved in the referendum or if they are used to complete a state research project. If a state research project is completed, companies in the Free Market Economy may also participate in the IC.

In IC, all the companies involved are equally liable and not the taxpayer. In an IC, all participating companies must pay their entire annual amount for research and development into a joint company account at People's Bank[84] and use it to pay for the IC's research and development.

The Innovation Agency organises paid professionals from the universities and institutes as employees and offers them to the IC until the participating companies have been able to hire enough of their own professionals. Funds, test labs and all research results are shared.

The IC may exist for no longer than the duration of the patent rights, i.e. a maximum of 20 years. The Antitrust Agency can dissolve it prematurely as soon as market entry has been achieved and 120% of the investment costs have been collected.

9.10.2 Export oligopoly

Domestic companies in the Social Market Economy and Planned Economy can apply to the Antitrust Agency for an export oligopoly if they want to jointly bring an innovative product to the world market but would not be able to do so alone. The innovation auditors check whether the conditions are met.

84Ministry of Finance - 11 People's Bank

9.10.3 Environmental innovations[85]

Innovations to protect the environment and health are financed through the prices of the products that cause damage. The price surcharge must be paid into the Innovation Fund. The price premium depends on whether a protective innovation already exists or not. If a protective innovation is already in place, the price of the innovation is used and how long it will take for the number of units of the damaging products sold to be sufficient to pay for the price of the innovation through the saved price surcharges. The health auditors monitor the time horizon until the innovation is imported during the audit. In voting with the Ministries of Health, Education and Labour, the Ministry of Innovation can enact laws that specify a time period until the innovations must be introduced.

For example, a company that previously obtained its electricity from oil, gas or coal would switch to electricity generated in-house. This would be feasible without additional costs and would also reduce costs in the medium term. The current price of electricity on the electricity exchange is about 30 cents per kilowatt hour (kWh).[86] The operating costs for a solar system are about 22.5 cents per kWh.[87] If the company were to consume 8500 kWh per year and pay 7.5 cents per kWh, the system would be paid off after 26 years if the system takes up 50 square metres of roof space and costs 16,500 euros.[88] This example shows that it makes more sense to use the protective technology immediately and to finance it with a loan that is paid off. The loan amount is financed by adding an additional fee to the electricity price, which goes into the pot for renewable energies in the Innovation Fund. The amount of the fee corresponds to the amount of the construction costs for renewable energies in the current year.

An interest-free loan is granted for the construction costs of innovation goods via the Innovation Fund. The interest is replaced by the fact that more revenue is generated for the

---

85§190.8 Environmental protection
86http://bricklebrit.com/stromboerse_leipzig.html
87200 euros per year are estimated for meter rental, insurance and cleaning for a system that supplies 4500kWh.
88https://www.haus.de/smart-home/solaranlage-kosten#a-164241-photovoltaik-kosten-und-ertrge

same costs or the same revenue is generated for lower costs. Because environmentally damaging products must include high disposal costs in the price, the price falls as soon as the environmental damage has been removed through innovation. The interest-free loan is granted through the Innovation Fund to all companies that introduce environmental innovations. The Ministry of Innovation collects the money through government bonds and pays interest on the bonds through the rising tax revenues due to falling costs or prices and rising profits or savings.

Companies that do not wish to take out a loan can also have the amount paid out after an audit by the Company Auditing Agency and then pay 3% of their profits into the Innovation Fund each year.

9.11 Financing

Funding for the review, granting and promotion of innovations is provided through fees and financial products of the Ideas Stock Exchange. For start-up financing, the Innovation Fund can be used or an Innovation Enterprise can be founded in the Planned Economy. For environmental innovations and state research projects, domestic guarantees are assumed or loans are granted directly by the People's Bank.

Funding for an innovation can be provided in full or only in part with the help of the Ministry of Innovation. The Innovation Agency's field service is responsible for any advice and motions for aid.

The prices for the examination, granting and maintenance of industrial property rights are to cover the entire costs incurred for this at the Patent Office, the Company Auditing Agency and the Innovation Agency and bring in 10% as a profit mark-up for the Ministry of Innovation. The costs of innovation auditors as part of the regular Company Auditing Agency audit of companies are determined by the Ministry of Labour. However, citizens can influence prices through the budget vote .[89]

The Ideas Stock Exchange financial products offer People's Bank

---

89 Ministry of Finance - 9.5 Budget vote

customers the opportunity to invest savings in innovations that are invented and marketed inland. The risk is reduced by the innovation auditors.[90] Investors can find out which inventions are popular via the challenge search engine and the search function. To do this, the directories for research, ideas, success models and the Innovation Database, are searched for profiles with the best ratings from Company Auditing Agency users and auditors.

9.11.1 Ideas Stock Exchange

The Ideas Stock Exchange is operated by People's Bank, but the requirements for investment products are determined by the Ministry of Innovation.[91] Innovations from the Ideas Directory and Innovation Database that have been reviewed by the innovation auditors can be traded on the Ideas Stock Exchange. Innovation bonds, innovation shares, licence participation shares, product shares can be bought or sold on the Ideas Stock Exchange.[92] Its main purpose is to facilitate financing and marketing for inventors or innovative companies, but also to offer investors a return on their investment. The return is generated by the monetary advantages of an innovation compared to its original state.

The Ideas Stock Exchange also offers funds for innovation bonds, innovation shares, licence participation shares, product shares and research costs. The Innovation Fund is a closed fund into which only involved companies or licensors pay and receive money after the Innovation Agency and innovation auditors agree.

---

90 Ministry of Finance - 11.10 Risk
91 Ministry of Finance - 11.9 Ideas Stock Exchange
92 Ministry of Finance - 11.9.1 Innovation Bonds, 11.9.2 Innovation Shares, 11.9.3 Licence Share, 11.9.5 Product Shares

9.11.1.1 Innovation Fund[93]

The Innovation Fund is a fund paid into by companies and inventors with successful innovations that once received money from the fund. All owners of a verified industrial property right can apply for money to market it. From this fund, money is paid out to domestic inventors per application to finance the development of a protected invention including prototypes and the establishment of a company as well as the purchase of special machines and personnel. Existing companies that want to import innovations but need money for new machines or skilled workers can also apply for money from this fund.

The Ministry of Innovation administers the Innovation Fund and, with the help of the Innovation Agency and the innovation auditors of the Company Auditing Agency, checks whether and how much money is paid out.[94]

Companies that were founded with money from the Fund and only market the industrial property right pay 5% of their annual profits into the Innovation Fund from the fifth business year onwards. Companies that were able to import the innovation with money from the Fund and for which the industrial property right only accounts for a share of the turnover only pay 5% of this share of the profits annually into the Innovation Fund. The more successful innovations are, the more money is available to finance new innovations.

In voting with the depositors, the Innovation Agency can set up an open pot and several closed pots. The open pot offers funds for any inventions. A closed pot ties the allocation of money to specific invention profiles or state research projects with economic urgency.

The Innovation Fund is invested at People's Bank and is listed on the Ideas Stock Exchange. This allows inventors to see which companies are involved in the fund and how much a particular invention is worth to the companies.

93§183,5,10 Research and innovation
94Ministry of Labour - 20.7.5.4.1 Approval of funds from the Innovation Fund

10 People's Innovation Company[95]

People's Innovation Companies are companies established and managed by the Ministry of Innovation. In essence, the management of the company is about producing and selling patented innovations that benefit the common good and the national economy until the industrial property right expires, i.e. a maximum of 20 years. If the invention is difficult to imitate, no patent is applied for, but it is produced secretly. If the invention is easy to imitate, a patent application is filed worldwide.

Due to the uniqueness and a high global demand, the profits should be as high as possible, replacing tax money. In this way, they benefit the domestic economy. People's Innovation Company's products should serve the common good because they are able to raise the standard of living of a population that uses them.

Because People's Innovation Companies have these two characteristics, they are subject to special conditions. They are the only companies that are allowed to be active in all economic forms. This means both the cooperation with companies of all economic forms in an innovation community and the buying and selling of goods and services in all economic forms and worldwide.

Apart from these exceptions, the rules for Social Market Economy joint-stock companies apply to People's Innovation Company.[96] The same rules apply to employees as to all other state employees. The Ministry of Education provides the necessary professionals until suitable staff is found and supplies the necessary expertise from its institutes. The Ministry of Infrastructure provides for the construction and maintenance of the buildings. The Ministry of Labour, through the auditors and advisors of the Company Auditing Agency, supports People's Innovation Company from its establishment through administration and management to privatisation or closure.[97]

Once a People's Innovation Company is established by the Minister of Innovation, the Company Auditing Agency's

---

95 §199.2 State enterprises
96 Ministry of Social Market Economy - 13.4 Joint-stock companies
97 Ministry of Labour - 7 People's Innovation Company

business consultants provide guidance during the establishment phase. When 110% of all expenses have been earned and paid back to the Ministry of Innovation, the start-up phase ends. Then the inventors of the patent and the employees of the People's Innovation Company jointly elect their managing director in a direct election of persons. As soon as the quorum of 50% of the employees and inventors is fulfilled, a new election takes place.

The people can decide in a voting to keep People's Innovation Company in state ownership for more than 20 years, to sell it earlier or to elect the managing director in a nationwide election of persons.

10.1 Initiators[98]

Any resident can submit an idea for a People's Innovation Company to the Innovation Agency. The Innovation Agency conducts a special search for new People's Innovation Companies with the help of various agencies. The first is the Patent Office, whose auditors forward promising newly filed patent applications. The second body is the People's Innovation Company Think Tanks, which have already successfully commercialised their innovation in a small company or filed a patent application. Third place is the innovation auditors of the Company Auditing Agency who find promising companies or patentable or already patented innovations when auditing new companies to be founded. Fourth are other ministries that promote inventive activity in their institutions. The Ministry of Education does this in its educational institutions and institutes.[99] Teachers of minors are encouraged to report inventions directly to the Innovation Agency that could become People's Innovation Companies. The Ministry of Planned Economy does this in the Social Villages through Innovation Workshops, Planned Enterprises, Innovation Enterprises, Research Institutions and in luxury

---

98 §154.4 Tax reduction, §183.2 Research and innovation: BV Art. 64
99 Ministry of Education - 4.10 Education through research, 11.7.2 State research institutes

supply.[100] Fifth are Social Market Economy and Planned Economy companies that are world leaders with an innovative product or service.

The Innovation Agency then makes the inventors or entrepreneurs a non-binding offer to turn their invention or company into a People's Innovation Company. If they vote in favour, the Innovation Agency creates an invention profile with the inventors, if this has not already been done. This invention profile is then submitted to the People's Innovation Company Commission for review. If the People's Innovation Company Commission agrees, this is noted on the invention profile. In that case, the inventor(s) will be offered exclusive production under licence, including voting rights in the management of the company. The inventor(s) can decide whether to accept the offer. If the inventor agrees, the production is immediately financed and carried out. From then on, the entire procedural costs for the patent are borne by the Ministry of Innovation.

10.2 People's Innovation Company Commission

The People's Innovation Company Commission audits patents for their ability to be marketed in a People's Innovation Company. Members of this commission are the Ministers of Innovation, Education, Labour, Planned Economy, Social Market Economy and Health. In addition to the ministers, participants include experts from their ministries who can assess the patent and its market value. All members of the commission are bound by professional secrecy, especially when it comes to non-protected inventions. The aim of the commission is to examine the patent's likelihood of profit and its usefulness in terms of the environment and the common good.

Specialised personnel from the state colleges support the commission. There, the invention is examined for its profit potential and its marketing is worked out. Students can write performance records or their final thesis on whether and how

---

100 Ministry of Planned Economy - 10.6 Innovation Enterprise, 10.5 Planned Enterprise, 10.9 Research and Development, 10.9.4 Innovation Workshops and Laboratories

the patent would be ready and marketable and how and where it could be marketed. Rarely does one student do this alone, but students from the affected subject areas share the work across the country. To this end, the commission advertises the necessary examination results as performance records in the Education Directory every semester. All students from all over the country may apply to the Examinations Office[101] with their previous marks for the examination. At least two complete proposals for development and marketing are submitted to the People's Innovation Company Commission. These two papers are then reviewed again at appropriate departments of the state universities and then submitted to the commission with a final evaluative rating.

Ultimately, the innovation minister decides whether his ministry can raise sufficient funds or whether tax money is needed to do so. If tax money is to be used outside the budget, a referendum must be held or wait until the next budget vote for the money to be approved by the people.

If the People's Innovation Company Commission agrees, the Minister of Innovation makes a binding offer to the inventor as to whether the state may establish a People's Innovation Company with his invention. If the inventor accepts the offer, a treaty is concluded between the Minister of Innovation and the inventor to establish a People's Innovation Company.

10.3 Building project[102]

All construction and equipment projects are planned, audited[103] , voted on and built by those responsible and affected. From the beginning, the future employees are involved in the projects as far as possible. People's Innovation Company construction should take place primarily in areas of the country that offer few jobs but plenty of housing.

---

101 Ministry of Education - 4.6 Examinations Office
102 §212.4 Structural policy
103 Ministry of Labour - 20.7.4 Technical auditor

10.3.1 Planning

The Construction Team's architectural office[104] is responsible for the planning together with the responsible deputy minister of the Ministry of Innovation. The Innovation Agency can request state employees from all ministries who work in similar areas or lend out specialists to create a professional and efficient working environment. In the planning process, the business consultants and all auditors of the Company Auditing Agency advise the responsible deputy minister.

10.3.2 Audit

The Company Auditing Agency checks with the economic auditors of Social Market Economy[105] whether the costs for construction and equipment are professional and efficient or whether there are cheaper or more efficient alternatives. If the audit is negative, the planning must be adapted to the requirements of the audit reports and a newly audit must take place. If the audit is positive, all participants involved vote on the building project. The Minister of Innovation has the right of veto and can thus approve or reject a construction despite opposing votes.

10.3.3 Voting

To prepare and hold the voting, a mobile show[106] will be held on Government Television[107] . If possible, the event will be held at or near the future construction site so that local residents can easily attend the event. All potential workers are searched through the Labour Directory and invited to the event. All viewers will have the opportunity to vote via their People's Computer.

In a presentation, the finished construction measures are shown digitally animated so that the object can be viewed

---

104 Ministry of Infrastructure - 5.8 Construction Team
105 Ministry of Labour - 20.7.3 Economic auditor
106 Ministry of Media Affairs - 5.7.2 Mobile Show
107 Ministry of Media - 7 Government Television

from a bird's eye view and from a first-person perspective. Trade secrets are shown in a blurred way.

All participants involved so far are present at the event. Namely, the Minister of Innovation, participants from the Ministry of Innovation, initiators, architects from the Construction Team and auditors from the Company Auditing Agency. If at least 60% of the people vote against the project, the People's Innovation Company will not be built at all. If at least 75% of the affected citizens at the possible construction site vote against, the People's Innovation Company will be built elsewhere.

10.3.4 Construction

The construction site is determined jointly by the responsible ministry and the affected citizens. Affected citizens are all residents within a radius of 500 metres or all residents of the municipality in which the construction site is located. Preferably, construction sites are lands owned by the state. All necessary buildings are constructed and equipped by the Construction Team.

All equipment, materials, machinery or services are purchased from companies in the Planned Economy and Social Market Economy and ordered through the Procurement Office[108] . Only if a good or service is not available in the Social Market Economy or Planned Economy, it is purchased on the world market at the best price-performance ratio.

10.3.5 People's Innovation Company Fund

Funding for the construction and equipping of new People's Innovation Companies is provided by the People's Innovation Company Fund. This fund is fed by deposits from existing People's Innovation Company and a trading fee for People's Innovation Company product shares. Existing People's Innovation Company pay 1% of their profits into the fund each year. 1% of the price paid for a People's Innovation Company

---

108Ministry of Labour - 6 Procurement Office

product share traded on the People's Stock Exchange goes into the fund. If a People's Innovation Company is privatised, part of the proceeds go into the People's Innovation Company Fund. If a People's Innovation Company is closed due to excessive losses, the fund bears the losses. If the amounts from the fund are not sufficient, a loan is taken out. In order to repay the loan as quickly as possible, all People's Innovation Companies must pay 10% of their profits into the fund until the loan has been repaid. Tax money should only be used in exceptional cases and if the people agree to it by a majority. The People's Innovation Company Fund is part of the budget vote and can distribute an amount to the next year's national budget if the people determine that it is overfilled.

10.4 Corporate governance

People's Innovation Companies are managed in a similar way to Social Market Economy joint-stock companies[109] . People's Innovation Company do not incur business taxes. Managing directors of a People's Innovation Company are directly elected and campaign with a programme for corporate governance. All employees, shareholders, inventors and, if required by a veto quorum, all nationals are eligible to vote. As soon as a deselection quorum of 50% of the group of persons consisting of employees, shareholders and inventors or 30% of the population votes for the deselection of the managing director, a new election is held. Like a minister, a managing director can make decisions independently or involve employees, shareholders and inventors in the decision-making process. At the annual general meeting, managing directors present their programme for the coming year and vote on it together with staff, shareholders and inventors. Annual general meetings are broadcast and moderated by Government Television.

---

109 Ministry of Social Market Economy - 13.4 Joint-stock companies

10.4.1 Industrial communities

People's Innovation Companies establish industrial communities for the purpose of innovation networking with domestic companies from all economic forms that have the most experience with needed technologies. From these companies, skilled workers are lent out, production steps are taken over, intermediate products or services are purchased or research laboratories are jointly operated. Domestic entrepreneurs are all companies that are majority-owned by domestic citizens, produce domestically and are domestically headquartered in one of the four economic forms. Work for a People's Innovation Company may not be done abroad or by foreigners. Companies that are in an industrial community with a People's Innovation Company pay only 50% of their business tax.

10.4.2 Procurement

Raw materials for production are sourced through the Procurement Office, which is sworn to secrecy so that patents, secret work processes or ingredients are not made public. In addition, People's Innovation Company have their own purchasing departments for special requirements.

10.4.3 Profit maximisation[110]

New and innovative products should be produced that can generate monopoly profits. Monopoly profits mean choosing the price and quantity produced at the level where the Cournot point is. Customer demand corresponds to the price-sales function, which states that at a higher price, a smaller quantity is sold. Marginal revenue is the revenue that results from the sale of one more product. Marginal cost is the cost of producing one more unit. The profit is the revenue (price times quantity) minus the costs (material and labour). In the following figure, the price on the Y-axis is in € and the quantity on the X-axis is in loaves. The profits of the baker

---

110 §183.2 Research and innovation: BV Art. 64

who sells 75 loaves at €2.40 each and has a cost of €1.50 per loaf is (2.4 times 75 minus 1.5 times 75) €67.5.

People's Innovation Companies that sell to their own state have to reduce the price until the price reaches the level where demand intersects marginal cost. In the example below, this would be about 2 euros per loaf. Depending on the assessment of the market situation, the managing director can reduce the price to a maximum of 10% above the point where marginal revenue equals marginal cost. This is the cost-covering price plus 10% profit. In the example below, 1.50€ production costs for a loaf of bread, so 1.65€ as the minimum price. This allows pricing to be adjusted to buyer demand to sell larger or smaller quantities. For example, if a product is produced that every human can use and would buy if the price is not too high, the lowest price is chosen that still yields at least 10% profits, but sold in as large a quantity as possible. A product that, for example, only states or large corporations want to afford is sold at the highest price in small quantities.

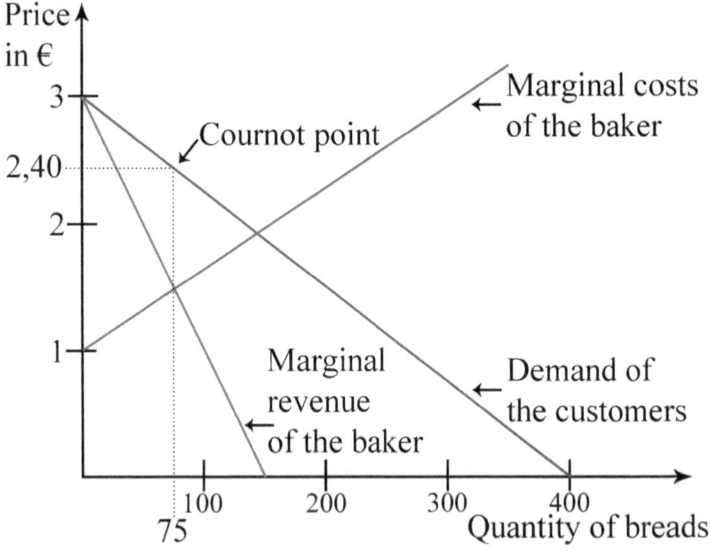

10.4.4 Losses

People's Innovation Company can make losses as long as they can offset them with profits in the next 3 years, otherwise the business is sold or closed. Once losses reach the level of the People's Innovation Company's construction costs, the People's Innovation Company is sold or closed. The proceeds from the sale go into the state budget.

As soon as losses occur that amount to 10% of the construction costs, the Company Auditing Agency's business consultants are called in.[111]

10.4.5 Ownership

Ownership rules regularisation of voting rights in governance votes. People's Innovation Companies are owned 15% by the inventor(s), 15% by the workforce and 70% by the people, represented by the Ministry of Innovation.

10.4.6 Profit distribution

During the start-up phase, 85% of the profits go to the Ministry of Innovation to generate the initial costs and pay for the services of the Company Auditing Agency. Inventors receive a share of the profits as a licence fee and do not have to contribute to the costs. Each year, 15% of the profits go to the inventor(s). Once the start-up phase is over, the 85% of the profits are distributed as follows. 10% is invested in research and development of the People's Innovation Company, 10% goes into the Research Cost Fund. 14% of the profits go to the staff as bonuses, in addition to wages, and each employee receives the same amount per working hour. 1% goes into the People's Innovation Company fund. The remaining 50% of the profits go into the national treasury and are available for distribution at the next budget vote.

10.4.7 VAT reduction[112]

As soon as the state has no debts and the people have no further use for the profits of the People's Innovation Company in the budget vote, the value added tax is reduced. The reduction

---

111 Ministry of Labour - 20.7.7 Business consultant
112 §154.3 Tax reduction

will be as high as the People's Innovation Company's paid-in profits and will be executed in the budget year after next. The reduction will be protected by the time lag in case profits or tax revenues are unexpectedly up to 100% lower in the coming year.

10.5 Participation in the financial market

People's Innovation Companies are considered joint-stock companies, but all shares must be owned by the inventors, employees and the Ministry of Innovation as the people's representative. The shares may only be sold when a People's Innovation Company is privatised. To raise money for investment, People's Innovation Companies can issue product bonds and product shares through the Ideas Stock Exchange.

10.5.1 People's Innovation Company product bonds

People's Innovation Companies are allowed to issue product bonds through the Ideas Stock Exchange, which can also be traded on the international financial market. Usually, bonds are agreed with a fixed term, a fixed amount and a fixed interest rate. What the company uses the money for is up to it. People's Innovation Company must use the money to produce their product as quickly and cheaply as possible while maintaining high quality. This is why these bonds are called product bonds.

If a People's Innovation Company's products are complex, lengthy and expensive to produce, product bonds may be issued to finance the production of the product. Once the product is sold, the bonds are paid out with the agreed interest. The term of the bond is linked to the manufacturing time of the product. If the production is delayed, the payout is delayed and vice versa. If a People's Innovation Company is closed, no more new product bonds are issued and the financed products are completed and sold. Investors are paid out with the sales proceeds. If the sales proceeds are lower, the investors suffer the loss. If the sales proceeds are higher, they flow into the state treasury.

10.5.2 People's Innovation Company product shares

People's Innovation Company can sell their products through the Ideas Stock Exchange so that consumers, usually expensive products, can rent them from shareholders. The rental price includes maintenance, if desired by the customer also operation and thus the costs for service-providing employees of the People's Innovation Company as well as the profits for the shareholders, which are paid out as dividends. Shareholders decide by majority vote at a shareholders' meeting how much the rent should be and how much the shares for operation and maintenance are. People's Innovation Company provide materials and personnel for operation and maintenance. However, shareholders can also have their product repaired or operated by other companies. People's Innovation Company product shares can also be traded on the international financial market.

10.6 Privatisation[113]

People's Innovation Companies are supposed to offer innovative and new goods and services as a product so that they alone can serve the market with them. As soon as competitors enter this monopoly market, the People's Innovation Company is converted into a Social Market Economy joint-stock company and the state sells half of its shares to the workforce. Each employee is entitled to the same number of shares per head. If one employee wants to buy fewer shares, other employees can buy more. If fewer employees in total want to buy shares than they are entitled to, the rest will be sold on the People's Stock Exchange along with the other half of the Ministry of Innovation's share package.

Once the inventions lose their monopoly status and competitors enter the market, the international price is lowered to 15% profits to increase the cost of entry for competitors. During

---

113§154.3 Tax reduction

this time, the shares held by the Ministry of Innovation are offered for sale to nationals and the workforce. The state is only supposed to make monopoly profits with People's Innovation Company to replace taxpayers' money. Once monopoly profits can no longer be generated, the People's Innovation Company will be privatised and marketed only to nationals through the People's Bank savings account. The issue price of a share is the current value of the People's Innovation Company divided by the number of shares issued. The number of shares is to be chosen so that the price of a share is equal to 10% of the current average monthly wage inland. The revenues from the share sales go into the People's Innovation Company Fund to finance the establishment of new People's Innovation Companies.

If the costs of a People's Innovation Company have not yet been recovered, the People's Innovation Company must continue to exist as long as it can show profits. Once the costs of the People's Innovation Company have been recovered and competitors are already in the market, the People's Innovation Company will be privatised. The revenues from the sale of the shares of the Ministry of Innovation should then at least cover the costs of the initial investments.

11 People's Innovation Company Think Tank

This People's Innovation Company serves to create jobs through inventions and gets therefore neither closed nor privatised. Here, the people are directly shown how People's Innovation Companies work, because the Ministry of Innovation produces many People's Innovation Companies, preferably more and more over time. They access locally available state property, especially during the creation phase, and use it to develop an innovation all the way to commercialisation.

11.1 Founding of a Think Tank

At least 5 persons can set up a Think Tank in the office of the Ministry of Innovation at the Town Hall and become part of the People's Innovation Company Think Tank. Anyone can have an idea that they would like to implement together with others. In order for the Think Tank to be workable, an order must be formed in which the ideas of the participants take their turn. In this way, at least 5 persons work on the realisation of an idea and not each person on his or her own idea. Ideas are written down on a list, which is worked through one by one. Anyone who comes to the Think Tank with a new idea is allowed to present it to all participants at the weekly meeting. The Think Tank decides whether the idea can be useful for the implementation of the currently discussed idea. If it is useful, it is immediately included in the work. If it is not useful, it goes to the bottom of the list. If, however, there is an idea on the list for which the new idea is useful, the new idea moves directly to the list position of the appropriate idea.

As more persons join the Think Tank through publicity and interest, groups can be formed. Once the first group has reached a size of 10 persons after the Think Tank has been established, 5 persons can split off and tackle the next idea on the list. The group of an idea remains until it could be implemented or had to be abandoned. Individual group members can leave or enter the group and swap with other groups.

Anyone new to a Think Tank can join any group and add their own new idea to the waiting list. If it is the turn of one's own idea in another group, one can switch to that group or wait until the group in which one is currently involved is ready.

11.2 State spaces

The People's Innovation Company Think Tank aims to provide spaces and materials that are owned by the state so that they can be used by all citizens to generate ideas. These spaces are existing vacant out-of-hours spaces, such as schools, sports fields, town halls or similar state institutions. Educational institutions already have special subject rooms for chemistry, art or similar. All these spaces are administered by the People's Innovation Company Think Tank for the period applied for. In some cases, training may be necessary for certain rooms with potential hazards, or there may be a requirement for an experienced specialist to be in the room.

Students, teachers or other working people who use the premises during normal opening hours may also register a Think Tank with the People's Innovation Company in order to innovate with other citizens at their workplace after hours. Pupils must be in school at least once during their school career while the school is being used by a Think Tank outside school hours.[114]

The timetable for all rooms within a selectable radius can be viewed in the town hall and in the Ideas Directory inventor groups. In the event of a shortage of space, allocation proceeds according to the following priority: firstly, pupils or persons using the state institution during its opening hours; secondly, private individuals in a club or Think Tank; and thirdly, companies or business unions in Planned Economy and Social Market Economy working with pupils and citizens in a research community.

11.3 Virtual Think Tank

In the virtual Think Tank, the group participants are not in the same place, but exchange ideas through texts, images, 3D animations or video or telephone conferences until the inventions are ready for the market or can become industrial property rights. They form an inventor group in the Ideas Directory and create an invention profile for it. The profile can also be created without content if they do not yet know what exactly they want to generate but need a digital platform.

11.4 Materials

Consumers must reorder used materials through the People's Innovation Company Think Tank. The People's Innovation Company Think Tank establishes contact with the responsible institution, arranges the order there and pays the material costs. In addition, members of a Think Tank can order materials through a Procurement Office catalogue.

---

114 Ministry of Education - 9.3.1.2 Research groups, 9.13 Opening hours, 9.15.1.2 School subject in the sixth learning year: Inventing

11.5 Use of the innovation workshops

Permanently administered by the People's Innovation Company Think Tank are the innovation workshops in the Social Villages.[115] The first pilot series can also be produced there because the other premises have to be used for other purposes due to the opening hours.
Citizens gain admission to these spaces by opening a Think Tank group in the Ideas Directory. Anyone gaining admission to the premises and institutions must be registered with their profile from the Persons Directory.

11.6 Financing

The People's Innovation Company Think Tank is financed by profitable innovations that have emerged from a Think Tank. Licensors or entrepreneurs are obliged to transfer 1% of the profits to the People's Innovation Company Think Tank via a surcharge on the business tax. The Ministry of Finance deducts the contributions. The corporate capital of the People's Innovation Think Tank is used to maintain premises with tools and materials, including the innovation workshops as factories of this People's Innovation Company. If companies find suitable inventions or employees, they have to pay a placement fee. The fee is a one-time 1% of the annual business tax, which is additionally collected by the Ministry of Finance from the company and forwarded to the People's Innovation Company.

11.7 Business cooperation

The People's Innovation Company supports the Think Tank in order to facilitate cooperation with companies. Surrounding companies should thus have the opportunity to draw on ideas from the population or find motivated skilled workers.

---

115 Ministry of Planned Economy - 10.9.4 Innovation workshops and laboratories

12 People's Innovation Company 3D printer halls

This People's Innovation Company builds and rents out halls that are specially designed for prototype construction and will replace conventional factories in the long term. Everything that cannot be grown through targeted gene design will then be produced there.

The conventional print-cast process is being replaced by oversized printing machines. The People's Innovation Company's printing programme works on an open-source basis, so that anyone in the world can program and print any product. In this way, anyone can design products and if there are enough buyers, many printing halls will be converted accordingly. One-offs and customised products are possible.

With their different spraying nozzles for plastics, metals, ceramics or stones, they create any plastic form. Influenced by form and function, different injection materials are produced in a wide variety of mixtures. With other attachments, metallic, alloyed or carbon-like yarn is interwoven, simultaneously bonded by a spray nozzle and trained into any kind of structure. Material can be removed by a milling machine. For this purpose, there is a titanium tip, a water jet or a laser. A wet-dry vacuum extracts overburden. With the laser, depending on the temperature, metal can also be welded or plastic can be welded.

Like many CNC milling machines and 3D printers, the machines operate on cranes and trolleys that can move horizontally and vertically through the hall. Print halls can quickly adapt or change their production. All that is needed is to change a different printer cartridge for the chemical mix, the print head for heat and tip, and the milling machine.

In this way, a wide variety of parts can be produced in the same hall, the size of which depends on the size of the hall. If there is an oversupply, it is possible to switch immediately to a shortage. Printing halls are created that can produce cars, furniture or prefabricated house components. To this end, the People's Innovation Company establishes printing hall industrial communities made up of domestic companies to build the new production facilities and train skilled workers. The chemical and mechanical engineering industries can rent these 3D printer halls.

Citizens with a tested patent are allowed to have a prototype built without waiting time and at cost-covering prices. Companies have to pay 20% more and plan for a waiting period until sufficient halls have been built. Companies that want to cushion their production peaks in the halls must take out insurance for this purpose, the funds from which are used to build halls and stockpile raw materials.

13 People's Innovation Company without more detailed description

The following People's Innovation Companies are only briefly listed. A more detailed description cannot be given here because these products are still patentable. They serve as seed money for the new policy system of dynamic media democracy.

1.	Infrastructurator „Turtle"
2.	Infrastructurator „Earthworm"
3.	Infrastructurator „Blue Whale"
4.	Infrastructurator „Pelican"
5.	Flying car
6.	Flying wing
7.	Space Elevator & Space Taxi

14 Symbol policy

The Ministry of Innovation is organising a symbolic policy to raise the profile of inventors in society. If the people do not agree with this expenditure, they can fix the amount in the budget vote. As a deputy for all other ministries, the Ministry of Innovation looks after decoration affairs. As soon as a minister deems a citizen worthy and wishes to bestow a decoration on him or her, he or she instructs the Ministry of Innovation to do so.

14.1 Good ideas day

On 13 August of each year, prize competitions are held for good ideas. In the week before, inventions are presented, similar to a film festival, and at the end of the week the most popular ideas are chosen by the citizens in a voting. To connect the whole thing with a prize competition, a lottery is offered. For each invention involved in the competition, raffle tickets can be bought during the week before. The day before the voting, ticket sales will end. All tickets cast for the three most popular inventions will end up in the lottery drum. The inventor with the most popular invention will draw a ticket from the lottery drum. The person who bought the ticket matching the ticket number wins. The revenues from the ticket sales form the prize money. The first-place inventor receives 25% of the revenues, 15% the second-place winner, 10% the third-place winner and 50% the winner with the matching ticket number.

14.2 Innovation competition

Every year, the Ministry of Innovation holds a competition for the best idea in the country. For this purpose, all industrial property rights in the Ideas Directory submitted inland are rated according to their generated turnover. This data comes from the Labour Directory and the People's Bank accounts for companies that have emerged from the inventions. Depending on how much turnover an invention has generated with licences or its own company, how many jobs it has created or secured, all this is evaluated by the Ministry of Innovation. The main prize goes to the invention that has performed best across all sectors. For each sector, all protected inventions that were on the market for one year and created the most turnover or jobs in the past year are also evaluated. The prize money for the most popular idea in an industry is the tax refund of the past 12 months for the company or inventor. The prize money for the most popular idea in the country, is the tax exemption for the next 3 years.

14.3 Statues for great domestic inventors

All previous successful domestic inventors receive a statue. The Minister of Innovation awards the prize to inventors whose ideas serve humanity because they produce technical and social progress. In addition, economic turnover and personal notoriety are used as selection criteria.

Anyone who creates an invention that makes him a billionaire within his lifetime or so famous that he is known worldwide receives a life-size statue made of stone. The inventor decides where inland it will be placed. If the statue is awarded after death, it is placed at the place of birth.

Anyone who creates an invention that has generated the value of the country's current annual budget since its market launch receives a 30-metre statue in the town of his or her birth. For example, Gottfried Daimler in Schorndorf and Karl Benz in Mühlburg would both receive a 30-metre statue as a community of inventors because they invented the car. Inside the statue is an exhibition about the invention and its triumphant progress. In the head is a viewing platform where you can look out of the eyes of the inventor. The entrance fees are used for preservation purposes, and any surplus always goes to all the inventor's descendants in equal shares.

In a capital city, the Field of Inventors is created next to a well-known landmark. There, the greatest domestic inventors are cast life-size in bronze and have a shield in their hands with the image of their invention. All inventors who also receive a 30-metre-high statue in their native city receive such a bronze statue. All inventors who have received a stone statue receive a life-size wooden statue in the city forest of the same capital city that has the bronze statues. This statue will weather over time and eventually be replaced by that of another inventor.

14.4 Museum for Innovation

A lying human is built as a house in survival size. The most popular inventions of past domestic citizens are exhibited and can be experienced as replicas. In the head of this human, there are permanent Think Tanks in two different laboratories, and in the belly there is an innovation workshop and 3D printer hall.

15 Switching to the new system

When switching to the new system, the Ministry of Innovation is first established. It takes over the Patent and Trademark Office, transforms it into the Patent Office and establishes the Innovation Agency. The Innovation Agency starts networking the domestic research institutions. All available funding for research and development is paid into the Research Cost Fund, the Innovation Fund and the People's Innovation Company Fund. The Innovation Agency builds the Research Directory and the Ideas Directory is built in cooperation with the Patent Office. An algorithm connects all existing funding programmes to the profiles in the Research Directory and Ideas Directory. The version of these directories can be handled via the Internet until the Intranet is available. The People's Innovation Company Think Tank is founded as the first People's Innovation Company.

15.1 Conversion of the old ministries

The following is a list of all the departments and units that are transferring to the Ministry of Innovation. If only the department or sub-department is named, all its units are transferred. If individual units are named, only these units are transferred. All departments and units not named are dropped. Existing staff adapt their tasks to the new requirements. The corresponding names of the units can usually be found as keywords in the running text.

15.1.1 Federal Ministry of Education and Research[116]

5 Research for digitalisation and innovation

7 Providing for the future - research for foundations and

116https://www.bmbf.de/upload_filestore/pub/orgplan.pdf Version: 03.06.2019

sustainable development
Large-scale facilities and basic research, universe and matter, international large-scale facilities (DESY, GSI), fusion, decommissioning of nuclear facilities
Sustainability, future provision, regional initiatives, mobility, energy, climate, biodiversity, marine and coastal research, bioeconomy, resources, circular research, geo research

15.1.2 Federal Ministry for Economic Affairs and Energy[117]

IV A Industry and mobility of the future

IV D Aerospace, defence and security industry, maritime industry

V Foreign trade policy
Trade fair policy, EXPO participations

VI C Innovation and technology policy

VI D Standardisation, safety
Technology transfer through standardisation and patents, fundamental issues of standardisation and patent policy, standardisation in ICT and secure internet architectures, product and plant safety

VII C SME and start-up financing, domestic guarantees

15.1.3 Federal Ministry of Food and Agriculture[118]

8 Rural development, digital innovation
Application of digitalisation in agriculture

117https://www.bmwi.de/Redaktion/DE/Downloads/M-O/organisationsplan-bmwi.pdf?__blob=publicationFile Version: 15.02.2019
118https://www.bmel.de/SharedDocs/Downloads/DE/_Ministerium/Organisationsplan.pdf;jsessionid=273685EE44CB88915756D4E49ACE8F5D.live841?__blob=publicationFile&v=8 As of May 2019.

15.1.4 Federal Ministry of Justice and Consumer Protection[119]

Z Administration of justice
Administrative matters of the DPMA, BPatG

I Civil law
Copyright, patent law

III Trade and Commercial Law
Copyright and publishing law, patent and inventor law, European Union patent and Unified Patent Court,
Trademark law, design law, combating product piracy

119https://www.bmjv.de/SharedDocs/Downloads/DE/Ministerium/Organisationsplan/Organisationsplan_DE.pdf;jsessionid=A807B5B1F5E FC74825E8B2A6508405BE.2_cid297?__blob=publicationFile&v=131
Viewed on: 14/05/2019

Contact form

Dear reader
If you would like to make what you have read come true, in whole or in part, together with other like-minded people, I offer you several possibilities with this contact form. Fill it out, tear out the page and send it by post to:
Andreas Seidl, P.O. Box 1206, 63488 Seligenstadt / Germany

Or send the details to:
Phone: 0049 1522 818 2243 (whatsapp, telegram, signal)
Email: andreas.seidl2022@web.de

Please mark with a cross:
O   I want to found a dynamic People's Party.
O   I want to donate money for implementation.
O   I want contacts with like-minded people in my area.

Forename: _______________________________________

Surname: _______________________________________

Please fill in only the contact option through which a reply should be made.

Street, house no.:   _______________________________

Postcode, city, country: _______________________________

Phone: _______________________________

Email address: _______________________________